GEOLOGY AND MINERALOGY RESEARCH DEVELOPMENTS

BENTONITE

CHARACTERISTICS, USES AND IMPLICATIONS FOR THE ENVIRONMENT

GEOLOGY AND MINERALOGY RESEARCH DEVELOPMENTS

Additional books in this series can be found on Nova's website under the Series tab.

Additional e-books in this series can be found on Nova's website under the e-book tab.

GEOLOGY AND MINERALOGY RESEARCH DEVELOPMENTS

BENTONITE

CHARACTERISTICS, USES AND IMPLICATIONS FOR THE ENVIRONMENT

AJAY KUMAR MISHRA
EDITOR

nova publishers
New York

Library of Congress Cataloging-in-Publication Data

Bentonite : characteristics, uses, and implications for the environment / Ajay Kumar Mishra (University of Johannesburg), editor.
pages cm. -- (Geology and mineralogy research developments)
Includes bibliographical references and index.
ISBN 978-1-63482-142-1 (hardcover)
1. Bentonite. 2. Clay. I. Mishra, Ajay Kumar, 1965- editor.
TN948.B4B375 2015
553.6'1--dc23

2015003404

Published by Nova Science Publishers, Inc. † New York

CONTENTS

Preface **vii**

About the Editor **ix**

Chapter 1 Clay and Clay-Supported Materials for Clean Energy Storage Applications **1**
Nicholas M. Musyoka and Henrietta W. Langmi

Chapter 2 Traditional and Modern Uses of Bentonite Clay and Implications on the Environment **19**
Derrick S. Dlamini

Chapter 3 Organo-Bentonite Adsorbents for Dyes and Other Organic Pollutants in Water **33**
Susmita Sen Gupta and Krishna G Bhattacharyya

Chapter 4 Bentonite and Its Composites: Removal of Heavy Metal Ions from Water **91**
Sadanand Pandey, Ajay K. Mishra and Shivani B. Mishra

Chapter 5 Applications of Bentonite and Its Role As a Base of Engineering Chemistry **127**
Tanvir Arfin

Chapter 6 Utility of Bentonite composite for Wastewater Treatment **137**
Rohit Bhatia, Bharti Arora, Jitender Gaur, Eun Ha Choi and Pankaj Attri

Chapter 7 Bentonite: A Versetile Clay **151**
Pradeep Pratap Singh and Ambika

Index **173**

PREFACE

Bentonite are usually found in deep geographical areas. They are produced from rock, fossils, weathering of volcanic ash and disposal of high-level of radioactive waste in many countries. Bentonite mainly consist of clay mineral such as montmorillinite and kaolinite. They mainly exist in cationic form of organic compounds such as aromatic or aliphatic hydrocarbon and are employed to modify high cations exchange capacity, higher structural integrity, thermal conductivity, high chemical buffering capacity, and stiffness. Bentonite is a layer of natural silicate clay mineral having numerous properties such as large specific surface area, great absorption and interlayer exchangeable cation. Bentonite consists of outstanding physico-chemical properties which make it applicable in the areas such as water treatment, medical, groundwater barrier, farming, and construction. They are well known to be one of the most versatile and inexpensive treatments.

Cleaner energy sources have been broadly studied for alternative energy sources in recent years. Solar energy, batteries and hydrogen have featured prominently amongst the most promising sources of cleaner energy. Clay materials being abundant, environmentally benign as well as having unique structural and physical properties have been considered as potentially low cost candidates for clean energy storage either in their pristine form and/or as support materials.

The occurrence of one octahedron-layer sandwiched between two tetrahedron-layers provides bentonite with vast applications in environmental problems. As a result of net negative charge associated with layers, counter positive ions have to be present in the interlayer to compensate the charge differences. Natural bentonite is abundantly available in many places and has gained importance as a low-cost adsorbent material for water applications. The adsorbing capacity of bentonite and its selectivity towards various pollutants could be modified substantially through the introduction of inorganic and organic substances into the clay structure. The hydrophilic character of the bentonite surface generally makes it less attractive as an adsorbent for non-ionic, non-polar organic pollutants in water.

The current book describes the bentonite characteristic, uses and implications for the environment in many areas including environmental science such as remediation, clean energy, water research and waste management. The researchers working in the area will benefit from the fundamental concepts and advanced approaches described in the book. The book is going to also be beneficial to the researchers who are working for their graduate and postgraduate degrees in the area of science and technology. The book also provides a platform for all researchers as it covers wide background for the recent literature, abbreviations, and summaries.

The book also covers a broader research area of chemistry, physics, environmental science, materials science, polymer science and engineering to present an interdisciplinary approach. The book contains the fundamental knowledge with the recent advancements for the research and development. The book also covers recent approaches and prospects about the current research and development of bentonite characteristic, uses and implications for the environment.

Ajay Kumar Mishra
Editor

About the Editor

Prof. Ajay Kumar Mishra is currently working as Professor at Nanotechnology and Water Sustainability Unit, College of Science, Engineering and Technology, University of South Africa, Florida Science Campus, 1709 Rooderport, Johannesburg, South Africa. He is also working as "Adjunct Professor" at Jiangsu University, China. Prof. Mishra has pursued PhD in Chemistry from Department of Chemistry, University of Delhi, Delhi, India. In 2006, he moved to the University of Free State, South Africa for Postdoctoral studies in the area of composites/nanocomposites. Later in 2009 Prof. Mishra has joined Department of Applied Chemistry as Senior Lecturer where he was promoted to Associate Professor in 2011. Prof. Mishra is currently group leader of the research area for the composites/nanocomposites, water research and bio-inorganic chemistry. He has hosted several visiting researchers/ scientists/postdocs in his group. Prof. Mishra has also developed a number of collaborations worldwide. His research contribution includes many publications in international journals. He has delivered a number of including Plenary/Keynote/Invited Lectures. For his outstanding research profile, he was awarded a number of awards. Prof. Mishra also served as Associate Editor as well as member of the editorial board of many international journals. He has edited several books by the renowned publishers. He has been reviewing a number of international journals and member of a number of scientific societies.

In: Bentonite
Editor: Ajay Kumar Mishra

ISBN: 978-1-63482-142-1

Chapter 1

CLAY AND CLAY-SUPPORTED MATERIALS FOR CLEAN ENERGY STORAGE APPLICATIONS

Nicholas M. Musyoka **and** ***Henrietta W. Langmi***
HySA Infrastructure Centre of Competence, Materials Science and Manufacturing, Council for Scientific and Industrial Research (CSIR), South Africa

ABSTRACT

The dwindling fossil energy reserves together with the need to find cleaner energy sources have intensified investigations for alternative energy sources. Solar energy, batteries and hydrogen have featured prominently amongst the most promising sources of cleaner energy. Among other challenges, the high cost of production together with limited reliable energy storage materials have contributed to the main obstacles for delayed the widespread switch toward cleaner energy sources. Clay materials being abundant, environmentally benign, as well as having unique structural and physical properties have over the recent years been considered as potentially low cost candidates for clean energy storage either in their pristine form and/or as support materials. This chapter presents a review of recent research work highlighting the use of clay and clay-supported materials for energy storage applications.

1. INTRODUCTION

Natural clays are formed by weathering of volcanic ash via a complicated process and are amongst the most abundant natural materials in the world [1]. They are also chemically inert and resistant to deterioration. The crystal structure of most clay minerals is mainly composed of sheets that are arranged in structural layers and hence reference to the terms 'sheet silicates' or 'phyllosilicates' [2-3]. These sheets are reported to be formed either by $Si0_4^{14-}$ tetrahedrons or by $[AlO,(OH),]^{6-}$octahedrons containing metal cations at the interiors of the tetrahedrons and octahedrons. Crystalline clay minerals are mostly classified into seven

[*] Corresponding author: E-mail: nmusyoka@csir.co.za.

groups [2]. These are; kaolinite and serpentine micas, vermiculite, smectites, pyrophyllite and talc, chlorites, and palygorskite and sepiolite group categories. These groupings fall into either type 1:1 “non-swelling” or type 2:1 “swelling” phyllosilicates. This classification depends on the number and combination of structural units [2]. For example, type 1:1 clays are composed of units of single sheets of silica tetrahedra between alumina octahedral sheets whereas type 2:1 clays minerals are formed when one central octahedral sheet is sandwiched between two tetrahedral sheets. The type of exchangeable ions in clay material has been reported to play a significant role in dictating its properties such as swelling capacity and hence influencing the type of application for the clay material [1,4]. For instance, certain cations in montmorillonite can be exchanged to other desirable cations such as by replacements of uni- and divalent cations by other di-, tri- or tetravalent cations for use as catalysts in the chemical industry [2].

Among the aforementioned groups of clays, the smectice group (such as montmorillonite, beidellite, nontronite and saponite) has found many industrial and domestic applications [2]. This is mainly due to their expanding/swelling properties [4]. For example, montmorillonitic clays have attracted applications as a drilling mud [1], water treatment [5] and also in energy storage [6-7], among many other applications. Since the objective of this book chapter is to briefly review the available information on applications of clay materials in energy storage application, readers interested in a more comprehensive understanding of fundamentals of clays and clay minerals, their properties and general clay applications, are referred to the ‘handbook of clay science’ edited by Bergaya and Lagaly [3]. This book serves as a good reference point for researchers starting or already working in the field of clay science.

2. Applications of Clay in Energy Storage

Among the many technological developments for energy storage, applications of clay materials have recently received significant attention. This review specifically focuses on applications of different clay-based energy storage technologies that range from thermal heat storage, storage of gases (i.e., either energy carriers or energy sources), use in batteries and supercapacitors as well as in other energy related uses such as in fuel cell technology.

2.1. Thermal Storage

The subject of thermal energy storage has been expanding over the last few decades especially due to the worldwide push towards renewable energy technologies [8-9]. The principle of thermal energy storage is based on the transfer of heat to a storage media during the charging period and later release during the discharging step. Various technological breakthrough have been achieved especially the ones that concern harnessing of solar generated thermal energy for later use. As reviewed by Xu et al. [9], there are three main thermal storage options; sensible heat storage, latent heat storage and chemical storage. Each of these technologies is reported to have its own advantages and disadvantages. Of the three available storage options, latent heat storage is reported to offer attractive features due to its relatively higher energy density [10]. Phase change materials (PCMs) such as inorganic systems, organic compounds and polymeric materials have been employed successfully as latent thermal energy carriers [10]. PCMs are known to store heat energy by undergoing a cycle of phase change induced by absorbing and releasing heat in the form of latent heat of fusion [8,9]. Other researchers have also prepared composites of PCMs by impregnation into porous structures. For instance, research by Sarier and co-workers [11] has shown that an organoclay composite prepared by intercalation of n-hexadecane with Na–montmorillonite, in a surfactant- assisted medium, possessed good heat storage and release capacity and could be utilized for thermal energy storage. Studies by Memon et al. [12] investigated the use of lauryl alcohol/Kaolin composites for thermal energy storage in buildings. Direct thermal energy storage in buildings has often been achieved by combining building materials such as clays with PCMs [12].

Other heat storage materials that employ adsorption onto solid material are also of interest. The most commonly investigated adsorbents are pristine zeolites and silica gel [13-14] although studies by Jänchen et al. [15] have also reported the use of modified zeolites and mesoporous materials for heat storage purposes. Of particular interest for this review, clay materials have been investigated in heat storage applications [6-7,16]. Most studies have reported that the best way to store thermal energy in clays is by hydrating their surface by using water vapor. This water vapour is mostly generated by solar energy or through industrial processes that produce waste heat. Clays are known to have the ability for absorbing about 7-10 quantities of water of its volume [6] and hence considered a promising material for thermal energy storage systems. Among the most widely used clays in heat storage are bentonites. These clays are attractive because they possess high heat of adsorption and are capable of hydrating and dehydrating without losing their structural stability [6-7,16]. The innate hygroscopic property of bentonite as well as its attractive heat of adsorption makes bentonite effective in the storage of solar and waste heat energy.

Among other properties, higher surface area of porous materials is considered an important property for enhanced thermal storage capacity [6-7,16]. Since clays are known to have relatively large specific surface area, they are considered desirable for energy storage applications. According to Konta [2], clay surface area mainly depends on its crystal chemistry in which case swelling clay minerals are reported to possess attractive external and internal surface areas. Other studies have also shown that sorption of water vapor by montmorillonite is also affected by the type of exchanged cation [4]. A review by Konta [2] highlighted that the differences in the degree of hydration and bonding forces of the interlayer Na^+ and Ca^{2+} cations is often the main factor that determines the difference in behavior of sodium and calcium bentonite. In this case sodium bentonite tends to be highly dispersive in water whereas calcium bentonite has a tendency to coagulate. These studies are also confirmed by Sadek and Mekhamer [6,7] who highlighted that type of cation, its size and the ionic charge plays a significant role in affecting the degree to which the clays stores thermal energy. A thorough study on effects of exchanged cation on the sorption of water vapors on montmorillonite clays was conducted by Chiou and Rutherford [4].

In underground heat storage systems, especially in the borehole thermal energy storage (BTES) option, a review by Givoni [17] reported that water-saturated clay and clay stones would be suitably used for underground long-term storage of solar energy. In natural conditions, other studies have shown that clay pans in semi-arid and arid desert such those as in areas around Queensland, Australia do possess a significant sources of sensible heat [18] and the resulting surface energy exchanges could be harnessed for ground heat storage systems. A field test report by Gabrielsson et al. [19], on heat storage in soft clay for temperatures up to 70 °C, highlighted that heat storage in soft clay could be feasible and also economically competitiveness. A study by Reuss et al. [20] also echoed similar recommendations and that better technical benefits could be derived from such high temperature systems.

2.2. Use in Gas Storage

Hydrogen is widely considered as an ideal energy carrier that could help address the issues of sustainability, carbon dioxide (CO_2) emission and energy security associated with today's fossil-based fuels. However, one of the major barriers to achieving a 'Hydrogen Economy' is the lack of suitable hydrogen storage systems. In this regard, hydrogen storage remains an intensively studied area of research [21]. Methane is also recognized as a promising energy carrier with a higher gravimetric energy density (50.0 MJ/kg) than gasoline (44.5 MJ/kg) [22]. Furthermore, methane has lower CO_2 emission in comparison with gasoline. When compared to coal, methane supplies about twice the energy of coal in terms of the amount of CO_2 release. However, a major challenge in the commercialization of methane as an alternative fuel is to develop safe and cost-efficient storage systems.

Today's established storage technologies for hydrogen and methane depend on high pressure or cryogenic conditions. These technologies are costly and require reinforced storage tanks, and in the case of cryogenic storage there is also an associated liquefaction penalty. Materials-based storage is attracting increasing attention and storage of these gases in porous materials is considered a promising option. There has been some interest in the use of clay minerals for gas storage either directly or indirectly because clays possess many attractive properties for their potential use in such applications [23]. Apart from their wide availability and low cost, their surfaces have intrinsic acid-base properties and adsorption sites can be tuned. They also have expandable frameworks, which can be exploited to tailor their specific surface area and porosity. Their interlayer spacing can be varied by incorporating different chemical species.

2.2.1. Hydrogen Storage

Several studies have reported on the use of clay or clay-derived materials in hydrogen storage. By intercalating lamellar clay minerals with chemical species such as silica, hybrid materials with enhanced surface areas and hydrogen adsorption properties can be generated. For instance, natural montmorillonite was intercalated with silica nanoparticles of diameter around 12 nm and the resulting intercalated clay had almost all the available surface exposed for hydrogen adsorption [24]. The separation between layers of the clay was at least 3.8 nm. Therefore, diffusion of hydrogen molecules to the interlayer was not restricted by the walls of the channels. At 77 K and atmospheric pressure, hydrogen adsorption increased from 0.12 for the pristine clay to 0.4 wt.% for the intercalated clay in corroboration with the increase in surface area.

Gil et al. [25] investigated alumina-pillared montmorillonite clays intercalated with solutions of hydrolysed aluminium, at various Al concentration/clay weight ratios, followed by calcination at different temperatures. The hydrogen adsorption capacity of the clays at 77 K was found to be strongly dependent on the microporous volume, and the clay with the lowest Al content (calcined at 773 K) exhibited the highest heterogeneity related to the presence of ultramicropores. Hydrogen adsorption properties could be enhanced by incorporating Pd nanoparticles in alumina-pillared clays. However, as the Pd content increased the microporosity and external surface area decreased due to the increase in size of the Pd nanoparticles, thereby influencing hydrogen uptake [26]. Jin et al. [27] modified natural diatomite by dispersing Pt and Pd nanoparticles on its surface. At 2.63 MPa and 298 K, hydrogen adsorption capacity of pristine diatomite reached 0.463 wt.%, while that of acid-thermally activated diatomite was 0.833 wt.% as a result of enhanced pore properties after activation. Upon inclusion of a small amount of Pt and Pd (approximately 0.5 wt.%), hydrogen adsorption capacities increased to 0.696 wt.% and 0.980 wt.%, respectively.

Clays have been used in the preparation of other hydrogen storage materials. Carbon nanotubes were prepared by the catalytic decomposition of acetylene over Ni/Cr hydrotalcite-type anionic clay catalyst [28]. High stability of the Ni nanoparticles prepared in situ enabled better dispersion of the active catalytic sites leading to the growth of a large amount of multiwalled carbon nanotubes (MWNTs). At 298 K and 80 bar, the hydrogen storage capacities of the MWNTs were reported to be 2.4 and 1.6 wt.% following acid and heat treatments, respectively. Hydrotalcite derived mixed oxides have the capability to generate highly dispersed metallic particles upon reduction [29]. By controlling the thermal decomposition of hydrotalcites mixed oxides of high specific surface area can be formed. Hydrogen gas can react chemically with the basic side of the hydrotalcite-derived metal oxides [30]. Salam et al. [30] synthesized nano-crystalline hydrotalcite-derived reduced mixed oxides containing magnesium, nickel and aluminium, and demonstrated successful nickel catalyzed reversible hydrogen adsorption near ambient conditions. Hydrogen uptake capacity of 3.9 wt.% was recorded, of which 1.9 wt.% was desorbed. The hydrogen adsorbed phase was confirmed as $MgH_2$and Mg_2NiH_4.

Clay-supported materials such as graphenes have been investigated for hydrogen storage. While the clay support (sepiolite) component did not adsorb any hydrogen, the graphene component exhibited hydrogen uptake [31]. In fact, at 77 K and 4 MPa, hydrogen adsorption capacity was 0.6 wt.% in relation to the total mass of the system (i.e., clay and graphene) and 1.7 wt.% in relation to the mass of graphene only. The high affinity between hydrogen and the clay-supported graphene as well as the existence of very narrow porosity were inferred from the shape of the isotherm. The isosteric heat of hydrogen adsorption was high (14.5 kJ mol^{-1}) and within the range of the optimum theoretical value reported for hydrogen storage at room temperature [32]. It was further shown that hydrogen adsorption was fully reversible. Clay-supported graphenes were also demonstrated to store hydrogen at room temperature although the capacities were lower than at 77 K [32].

Catalysts supported on clays have been employed in the generation of hydrogen from chemical hydrogen storage materials. Dai et al. [33] prepared Ru nanoparticles immobilized in montmorillonite and investigated its catalytic effect on the methanolysis reaction of the hydrogen storage material, ammonia borane. It was observed that the Ru component existed as intercalated Ru nanoclusters between the silicate layers and also as Ru nanoparticles deposited on the external surface of montmorillonite. The Ru catalyst supported on montmorillonite showed high performance and was highly effective in facilitating the methanolysis reaction of ammonia borane. The average hydrogen generation rate was 29 L min^{-1} g^{-1} (Ru). In addition, the catalyst exhibited high durability during cycling, retaining 95 % of its initial activity and ensuring 100 % conversion in its 20th cycle (Figure 1). Alkaline sodium borohydride ($NaBH_4$) solution is an attractive hydrogen storage medium. However, to enable a satisfactory release of hydrogen a catalyst is needed. Chamoun et al. [28] prepared clay-supported Co catalysts using three different clays. Two of the supported catalysts of which the supports mainly composed of kaolinite and illite, respectively, showed $NaBH_4$ conversions of 100 % at 60–80 °C and hydrogen generation rates up to about 31 L(H_2) min^{-1} g^{-1}(Co)). An attapulgite clay-supported cobalt-boride (Co-B) catalyst has also been used to generate hydrogen from $NaBH_4$ [34]. It was reported that the average hydrogen generation rate decreased slowly from 1.27 to 0.87 L min^{-1} g^{-1} Co-B from the first to the ninth cycle.

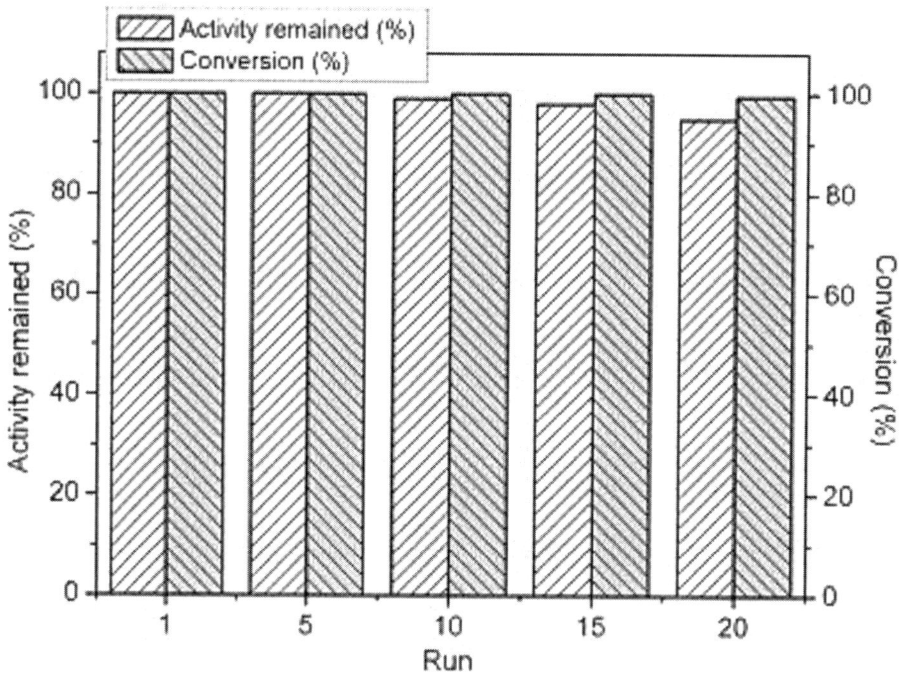

Figure 1. Cyclic performance of the Ru/montmorillonite catalyst (Ru/ammonia borane molar ratio = 0.02) in catalyzing the methanolysis of ammonia borane [33]. Copyright 2010 Professor T. Nejat Veziroglu.

2.2.2. Methane Storage

Studies of the adsorption of methane on clays are limited since it is believed that mineral matter, in essence, is non-adsorptive to methane [35-36]. However, storage of methane in clays has been investigated in the context of exploiting shale gas resources [37-38]. Shale gas presents an unconventional gas resource, which could potentially expand the worldwide energy supply. Shale gas comes from organic matter in shale and is trapped in shale formations. The dominant component in shale gas is methane. Shale gas may be stored in several forms i.e., as free, sorbed or dissolved gas [39]. Knowledge of gas storage is necessary to gain a better understanding of the reservoir and its long-term production of shale gas, in order to reduce exploration risks and determine economic feasibility [37,40].

The principal constituents of shale are organic matter and clay minerals, both of which play a part in methane adsorption. Lu et al. [41] carried out adsorption measurements on Devonian shale samples. A considerable amount of the natural gas in the Devonian shales in the Appalachian basin is stored as a condensed phase, which includes the adsorbed gas on surfaces and the dissolved gas in organic matter. In this study, 'adsorbed gas' referred to the condensed gas phase. A linear relationship was observed between adsorbed capacity and total organic content, implying that organic matter is one of the key factors responsible for adsorbed gas storage [41]. Nevertheless, shale samples with low organic matter content could still adsorb significant quantities of gas. It was further proposed that the clay mineral, illite, might play a role in adsorbed gas storage since Devonian shale consists of illite, carbonates and quartz and the latter two components do not adsorb methane. When the adsorption isotherms for illite at 37.8^{o}C and approximately 7 MPa were compared with those for Devonian shales it was observed that the adsorption capacity for illite is significant and therefore capable of contributing to the total gas storage in Devonian shales.

Cheng and Huang [42] experimentally studied the effect of adsorption process on the compositional variations in hydrocarbon gases migrating through clays (kaolinite and montmorillonite), and organic matter (coal, oil shale and activated-carbon). The experiments were performed under dry condition and low pressure, which was noted to be simple compared to a complex real system where gas retention and migration occur in the presence of fluid phases, and adsorption takes place in a multi-phase system at higher sub-surface pressures. It was reported that even though the adsorption of methane and wet gases on clays was less than that for organic matter, it was still significant. It was apparent that under the experimental conditions employed the larger internal surface area of montmorillonite did not substantially enhance its adsorption capacity of hydrocarbon gases.

In another study, Ross and Bustin [37] investigated the effect of shale composition and pore structure on methane storage potential of shale gas reservoirs. They reported that inorganic material has an effect on modal pore size, total porosity and sorption characteristics of shales. Methane sorption capacities of dry and moisture equilibrated illite, montmorillonite and kaolinite varied significantly. At 30°C and 6.0 MPa the methane storage capacities of illite (2.9 cm^3g^{-1}) and montmorillonite (2.1 cm^3g^{-1}) exceeded that of kaolinite (0.7 cm^3g^{-1}) on a dry basis, in agreement with the trend in micropore volume and surface area. While on a moisture-equilibrated basis, kaolinite (0.7 cm^3g^{-1}) adsorbed more methane than illite (0.4 cm^3g^{-1}) and montmorillonite (0.3 cm^3g^{-1}) as a result of the low moisture content of kaolinite relative to illite and montmorillonite.

High pressure adsorption of methane on Ca^{2+}-montmorillonite, kaolinite and illite was reported [38]. The measurements were performed at 60 °C and pressures up to 18.0 MPa. The clays exhibited high methane adsorption capacities and the maximum Langmuir adsorption capacities achieved were 6.01, 3.88 and 2.22 cm^3g^{-1} for montmorillonite, kaolinite and illite, respectively. Methane adsorption occurred only on the external surface for kaolinte and illite. However, in the case of montmorillonite methane adsorption occurred both on the external surface and in the interlayer space since the interlayer distance was larger than the size of a methane molecule. The interlayer space of montmorillonite was accessed by methane at low pressures with more methane molecules gaining access at high pressures. It was also demonstrated that the adsorption sites of the clay could be occupied by water leading to a decrease in methane adsorption capacity. From these studies it is clear that clay minerals play a significant role in methane adsorption of shale.

2.3. Use in Lithium Ion Batteries and Supercapacitors

With the growing need for cleaner, more efficient energy and storage the demand for lithium-ion (Li-ion) batteries has been increasing rapidly due to their unique advantages such as high gravimetric and volumetric energy densities, high power density, single-cell voltage, long cycle life and low self-discharge property. While Li-ion batteries have been shown to be the most efficient energy storage approach for a variety of portable devices such as cellular phones and laptops, their use in (hybrid) electric vehicles requires about two to five times more energy density than what today's lithium batteries can offer (150 Wh/kg) [43]. A Li-ion battery is a rechargeable battery normally made up of a carbonaceous anode (negative electrode) and a lithium transition metal oxide cathode (positive electrode) that are separated by a porous membrane immersed in a non-aqueous electrolyte. The separator prevents electrical contact between the electrodes while allowing the diffusion of Li-ions from cathode to anode during charging and discharging. Supercapacitors use a different energy storage mechanism from batteries that do not involve chemical reactions. A supercapacitor stores electrical energy in the electrical double layer that forms at the interface between an electrolytic solution and an electronic conductor. The main components of a supercapacitor are also electrodes (typically high surface area activated carbon), an electrolyte (aqueous organic solution or ionic liquid) and a separator, which prevents contacts between the electrodes. Supercapacitors offer several advantages over Li-ion batteries such as longer lifetime (can last for half-million cycles or more), high charge and discharge rates, and high power density. However, their main drawback is the low gravimetric density compared to Li-ion batteries. Research efforts are being devoted to improve the components of Li-ion batteries and supercapacitors so as to produce safe, low cost, high performance devices that can satisfy energy storage needs.

2.3.1. Electrolytes

Interests in the development of polymer–clay nanocomposites for electrochemical applications began after Ruiz-Hitzky and Aranda [44] reported the intercalation of poly(ethylene oxide) in lithium-smectites, which gave rise to Li single-ion conductors. In this regard, the application of polymer-clay nanocomposites is in the development of electrolyte and electrode components for energy storage devices, primarily rechargeable Li-batteries and to some extent supercapacitors. As electrolytes, it is a requirement for the polymer-clay nanocomposite to display good ionic conductivity at room temperature, and also to possess good thermal and mechanical properties. Polymer gel electrolytes based on polymer–clay nanocomposite show promise for application as electrolytes in Li- ion batteries. Shubha et al. [45] fabricated polymer gel electrolytes (PGEs) based on electrospun poly(vinylidenefluoride-co-hexafluoropropylene) {P(VdF-co-HFP)} incorporated with layered nanoclay. The electrochemical studies demonstrated that inclusion of layered nanoclay in the polymer matrix considerably enhanced the ionic conductivity and compatibility with lithium electrodes. Highest room temperature ionic conductivity of 5.5×10^{-3} S cm^{-1} was reported for P(VdF-*co*-HFP)–$clay_{(1\%)}$–PGE. The interfacial resistance was substantially reduced by a faster stabilization of the passivation layer due to the addition of clay, thereby resulting in better compatibility with the lithium electrode. The membranes with nanoclay exhibited high porosity, electrolyte uptake, thermal stability and enhanced electrochemical properties. The effect of clay on the membrane morphology, ionic conductivity and electrochemical properties of the PGEs in Li-ion rechargeable batteries was also studied by Prasanth et al. [46]. Inclusion of 2 wt.% nano-clay in the poly(vinylidenefluoride) matrix substantially improved its ionic conductivity from 0.78 x 10^{-3} to 3.08 x 10^{-3} S cm^{-1} at room temperature. Here also, the interfacial resistance with lithium metal electrodes was considerably reduced due to the presence of nanoclay in the PGEs and a stable cycle performance of the Li-ion half-cell was reported.

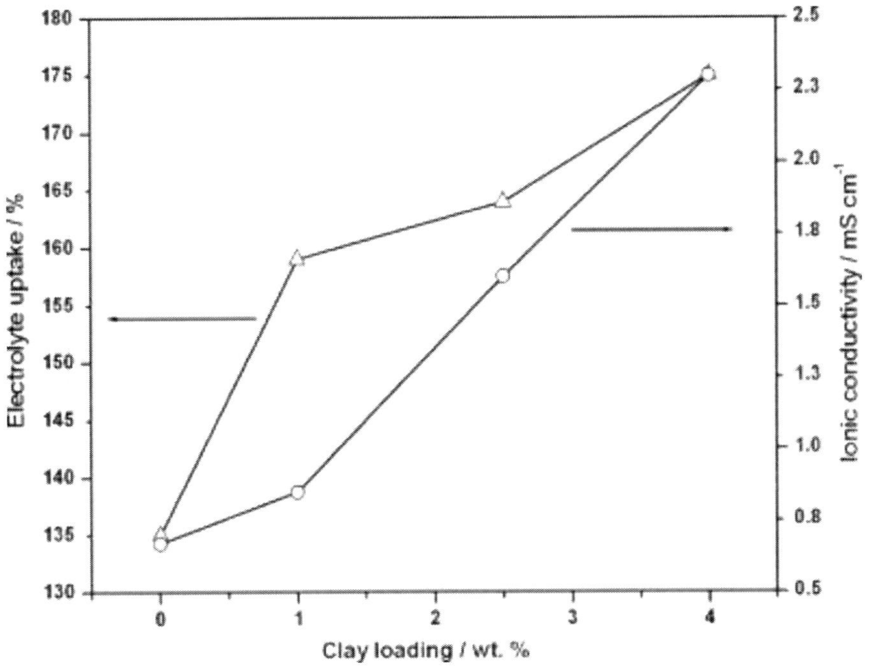

Figure 2. Ionic conductivity (right ordinate) and electrolyte uptake (left ordinate) of nanocomposite gel polymer electrolytes as function of organophilic MMT [47]. Copyright 2010 Elsevier B.V.

Deka and Kumar [47] investigated nanocomposite polymer electrolytes based on intercalation of poly(vinylidene fluoride) polymer into organically modified montmorillonite clay. The ionic conductivity of the resulting nanocomposite gel polymer electrolytes increased as the clay loading was increased, and reached a maximum value of 2.3 x10^{-3} S cm^{-1} at room temperature for a 4 wt.% clay loading (Figure 2). This behaviour was ascribed to the higher uptake of liquid electrolytes by the nanocomposites and the dissolution of more lithium salt due to the presence of the high dielectric constant clay layers. At this loading the nanocomposite gel polymer electrolyte exhibited superior electrochemical and interfacial properties over a clay-free electrolyte system. Poly(vinylidene fluoride)-hexafluoropropylene membranes containing various kinds of clay have been shown to possess an increased membrane modulus compared to the pristine Poly(vinylidene fluoride)-hexafluoropropylene membrane [48]. Chen-Yang et al. [49] used organically modified montmorillonite as a filler to prepare a series of composite polymer electrolytes based on polyacrylonitrile (PAN) and $LiClO_4$. The incorporation of organoclay into the PAN/$LiClO_4$ polymer electrolyte made the composite a better polymer electrolyte than that without the clay. The addition of organoclay enabled the free movement of Li^+ ions in the polymer electrolytes and allowed dissolution in the PAN matrix, thereby increasing the concentration of the solvated Li^+. As such, the ionic conductivity of the as-prepared organoclay-containing polymer electrolyte was enhanced. In fact, the ionic conductivity of the PAN/$LiClO_4$/organoclay nanocomposite polymer electrolyte (with 7 wt.% of the organoclay) was more than seven times higher than that of the corresponding PAN/$LiClO_4$ polymer electrolyte. The as-prepared composite polymer electrolytes also exhibited excellent electrochemical stability and cyclability.

2.3.2. Electrodes

Polymer–clay nanocomposites developed for application as electrodes in batteries and supercapacitors are primarily intermediates for the preparation of porous carbons displaying electronic conductivity as well as lithium insertion capacity [50]. Carbonaceous materials are usually used as electrodes in batteries and supercapacitors. These carbon materials can be prepared using various approaches with the templating approach considered one of the best methods to produce conducting carbon with high surface area and porosity. Clays, among other porous substrates, can be used as templates in which molecular or polymeric organic precursors are adsorbed and subsequently pyrolized to obtain the templated-carbonaceous material, which can be further treated to remove the template. Alternatively, to synthesize carbonaceous materials, intermediate polymer–clay nanocomposites can firstly be prepared either by polymerizing previously adsorbed monomers (e.g., acrylonitrile, furfuryl alcohol, vinyl acetate and sucrose) on the clay or by directly incorporating polymers (e.g., polyfurfuryl alcohol, polyvinyl acetate and polyvinyl alcohol) in the clay. These intermediates are then thermally treated to obtain either highly oriented carbon–clay or disordered carbonaceous materials [50]. Subsequent acid treatment removes the silicate substrate and generates the carbonaceous material, which preserves good stacking of the graphite-like layers, although in some cases the layered structure is lost [51]. Duclaux et al. [52] prepared carbon by pyrolysis of organic precursors in the interlayer space of taeniolite. Although the carbons afforded suitable properties for lithium insertion, solvated lithium ions were easily trapped in the mesopores. Thus, inspite of the low specific surface area, the irreversible capacity was very high, rendering the carbons unsuitable for application in Li-ion batteries. However, carbons obtained from the clay-carbon nanocomposites at low temperature (700-850°C) showed promise as electrode materials for supercapacitors. Specific capacitance of 85 F g^{-1} was recorded on capacitor cells in aqueous electrolyte.

Fernández-Saavedra et al. [53] investigated the synthesis of polyacrylonitrile–sepiolite nanocomposites as precursors of carbon–sepiolite nanocomposites and carbon nanofibers and reported reversible lithium insertion, which could be relevant for application as anodes in rechargeable Li-batteries. It was reported that the carbon-sepiolite nanocomposite could have reversible Li insertion with a specific capacity of 406 mAh g^{-1}. Meanwhile, as expected, removal of the silicate to form free carbon nanofibers enhanced the Li insertion behaviour displaying a considerably higher specific capacity of 830 mAh g^{-1}. In another study, Fernández-Saavedra et al. [54] prepared nanostructured carbonaceous materials of various microstructures from nanocomposites containing polyacrylonitrile inside the nanosized pores of imogolite, sepiolite and porous alumina membrane templates. The nanostructured carbonaceous materials demonstrated high capacities for Li insertion. Carbon nanotubules derived from porous alumina membranes showed higher reversible specific capacities than nanostructured carbons derived from clays. The values for the carbon nanotubles were directly related to the pore size of the alumina membrane. The reversible Li insertion capacity for nanostructured carbon derived from imogolite was higher than for nanostructured carbon derived from sepiolite. This difference was attributed to several factors such as a higher surface area, microporosity, and electrical conductivity of imogolite. The nanostructures carbonaceous materials also displayed high specific capacitance values when used as electroactive materials in electrochemical double-layer capacitors. Those derived from imogolite and sepiolite exhibited capacitance values of 22.4 F g^{-1} and 50.2 F g^{-1}, respectively, which were relatively lower than the values for carbon nanotubules derived from porous alumina membranes (62.7 F g^{-1} for 100 nm pore size alumina membranes and 46.2 F g^{-1} for 200 nm pore size ones).

Carbon-clay composites have also been examined as electrode components. Wang et al. [55] derived porous Si/C composites, comprising porous Si coated with a 15 nm thick amorphous layer of carbon, from diatomite – a natural clay mineral. The porous Si/C composites with various ratios of Si and C were investigated as anode materials for lithium-ion batteries. It was shown that the porous Si/C composite containing 33 % carbon displayed the highest reversible capacity (1628 mAh g^{-1}) at the first cycle, with excellent capacity retention in subsequent cycles. In addition, the porous Si/C composites showed excellent rate performance at high current densities. The findings illustrate that the clay mineral, diatomite could be considered as a promising raw material to produce Si/C composites for use as anode materials in Li-ion batteries.

Functionalized conducting carbonaceous materials were prepared from sucrose and the clay mineral sepiolite [56]. The resulting graphene-like sepiolite nanocomposites were evaluated as electrode materials for rechargeable Li-ion batteries, and were shown to exhibit better cyclability and Li-insertion properties than the nanostructured carbon without the silicate component.

This difference was attributed to the special texture of the carbon-sepiolite nanocomposites exhibiting high specific surface area and porosity as well as a high electrical conductivity that favoured lithium insertion. The removal of the silicate template by acid resulted in deagglomeration of carbon particles with a concomitant increase in the specific surface area, but decreasing the electrochemical lithium insertion capacity as a result of the electrolyte aggressively penetrating the electrode material. It was highlighted that the graphene-like sepiolite materials derived from sucrose–sepiolite precursors are very efficient materials for electrodes of rechargeable Li-batteries, and are the first carbon–clay material suitable for this application not warranting elimination of the silicate template component. Ruiz-García et al. [31] employed an environmentally friendly protocol to prepare clay-supported graphenes from caramel and sepiolite. The materials synthesized in this manner exhibited good electrical conductivity and high specific surface area afforded by the presence of the sp^2 carbon in graphenes and the silicate backbone in clay, respectively. The clay-supported graphenes could act as electrode for lithium batteries without the need for conductive additives as normally required. Model lithium-rechargeable batteries were tested, where the electrochemical cells contained supported graphenes as anode materials and lithium metal as the cathode.

It showed specific capacity values close to 400 mAh g^{-1} and improved cyclability behavior. The clay-supported graphenes were also tested in supercapacitors revealing poor specific capacity values around 30 F g^{-1}, possibly because the device set-up was not optimized.

2.4. Other Energy Related Uses

2.4.1. Fuel Cells

A fuel cell is an electrochemical conversion device with a continuous delivery of fuel (such as hydrogen or methanol) and an oxidant (such as oxygen or air). It converts the chemical energy of the fuel into electricity by means of a chemical reaction with the oxidant. The main components of a fuel cell are an anode, a cathode and an electrolyte permitting the movement of charges between the two sides of the fuel cell. The major challenges to overcome in fuel cells research are improving efficiency, durability and cost. Materials research on the constituents of a fuel cell is of paramount importance. The proton electrolyte membrane (PEM) fuel cell is considered the most promising for transportation applications. Its operating temperature is near 100°C with near-zero or zero emissions of environmental pollutants. A major hurdle facing the commercialization of direct methanol fuel cells is the methanol crossover, which becomes considerable at a high methanol concentration. When methanol crossover occurs, the methanol undergoes a reaction at the cathode and no electricity is generated, thereby decreasing the fuel usage and reducing the activity of the catalyst at the cathode.

Polymer-clay nanocomposites have been investigated for the reduction of methanol migration rate across the membrane [57-60]. For instance, Felice et al. [58] evaluated the effect of Nafion-montmorillonite nanocomposite membranes on hydrogen and methanol crossover in a PEM fuel cell. It was reported that both the addition of clay to the membrane and the thickness of the clay-nanocomposite membrane were key parameters affecting crossover. The Nafion-clay nanocomposites exhibited substantially reduced permeability to both methanol and hydrogen when compared to a control membrane of the same thickness. However, the ionic conductivities of the Nafion-clay nanocomposites were lower than that of pristine Nafion membrane of similar thickness. Meanwhile, incorporation of methylpropanesulfonic acid-modified montmorillonite into Nafion matrices led to substantially reduced methanol crossover while retaining conductivity properties [59]. A fourfold proton conductivity enhancement was observed for sulfonated poly(ether ether ketone)-montmorillonite nanocomposite membrane after sulfonation of the montmorillonite [57]. A reduction in methanol crossover was also reported for the sulfonated poly(ether ether ketone)-sulfonated montmorillonite nanocomposites. Sulfonation of polysulfone as well as incorporation of Cloisite 30B as a filler in polysulfone-Cloisite 30B nanocomposites led to an increase of proton conductivity [61]. Generally, nanocomposite polymer electrolytes consisting of Nafion reinforced with smectite clays were reported to be stiffer, possessing high thermal stability, which are desirable properties for fuel cell applications [62]. Improvement in thermal stability of Nafion membrane was also achieved by the addition of carbon nanotubes rooted montmorillonite into the Nafion matrix [63]. It is worth noting that apart from their use as electrolyte membrane components, clays have also been examined as electrode components in fuel cells, for example, bifunctional graphitized carbon-supported Pt oxygen electrode with montmorillonite-assisted dispersion [64], (polypyrrole-montmorillonite nanocomposite electrode [65] and Pt-laponite clay-modified gold electrode [66].

2.4.2. CO_2 Capture

It is generally accepted that CO_2 emissions play a major role in global warming and associated climate change. Reducing anthropogenic CO_2 emissions is an essential requirement for mitigating the potential and realised effects of climate change. CO_2 is considered the most important among the greenhouse gases, accounting for about 64 % of increased greenhouse gas [67]. The world economies are heavily dependent on fossil fuels as a primary energy source, which upon combustion releases large amounts of CO_2. It is estimated that the energy demand and CO2 emissions will more than double by 2050 [68], thus mounting the pressure on governments to look for new ways to alleviate CO_2 emissions. As such, many countries have committed to stringent targets for reducing CO_2 emissions. Various technologies for CO_2 capturing are being developed worldwide and it is essential to aim towards achieving an effective, low-cost technology that is environmentally friendly and readily applicable.

Clays have been explored as potential candidates for CO_2 capture. For instance, Volzone et al. [69] investigated the intercalation of tetramethylammonium (TMA) and hexadecyl pyridinium (HDP) in two forms of montmorillonite for CO_2 adsorption. It was shown that the amount of CO_2 adsorbed increased after treatment with TMA but it decreased after intercalation with HDP cation, in accordance with the larger size of HDP compared to TMA. This was attributed to the fact that the smaller sized TMA do not form a continuous organic phase on the clay surfaces and in the interlayers, leaving the siloxane surfaces available for CO_2 retention. The CO_2 retention by TMA-montmorillonites was in the range 0.383-0.477 mmol g^{-1}. The influence of ion-exchange in montmorillonite on CO_2 retention was further investigated using the following ions: Na^+, K^+, H^+, Ca^{2+} and Al^{3+} [70]. The CO_2 adsorption by the cation-exchanged montmorillonites followed the order: H- > K- > Ca- > Al- > Na-montmorillonite. The adsorption of CO_2 was influenced by several factors including ionic radius of the exchangeable cations, the interlayer separation and surface area of the montmorillonite. CO_2 retention in acid treated smectites have also been reported [71]. An increase in CO_2 retention occurred after acid treatment. The acid trioctahedral smectite had higher CO_2 adsorption than acid dioctahedral smectite before and after thermal treatment up to 900 ^oC. The CO_2 adsorption values by acid smectites heated up to around 600 ^oC, decreased by about 10 and 20 % for acid trioctahedral and dioctahedral smectites, respectively. The chemical composition, structure and texture of the acid treated smectites were the factors that influenced gas retention.

The intercalation of montmorillonite with polyol dendrimers to produce organo-montmorillonites bearing weak base OH sites was reported [72]. The materials were demonstrated to reversibly capture CO_2, with high CO_2 retention values of 0.5–2.5 mmol g^{-1}, and regeneration was attained under slight heating up to 100 °C. The CO_2 retention capacity was related to the number of OH groups of the dendritic moiety, dendrimer content and the specific surface area. Below 100°C, CO_2 adsorption was higher than predicted by stoichiometry, revealing that the interactions between CO_2 and OH groups were only physical. Recently, Elkhalifah et al. [73] prepared several bentonite clay adsorbents modified by amines by the intercalation of protonated mono-, di- and triethanolamines into the interlayer space of bentonite for their use as adsorbents for CO_2 capture. The study showed that the CO_2 uptake capacities for all the bentonite clay adsorbents modified by amines increased by approximately 2.5 times when compared to unmodified bentonite (Figure 3). Furthermore, the CO_2 adsorption capacity for bentonite clay modified with monoethanolammonium cations was higher than that of bentonite clay modified with di- and triethanolammonium cations. This observation was attributed to the higher BET specific surface area of the clay modified with monoethanolammonium cations.

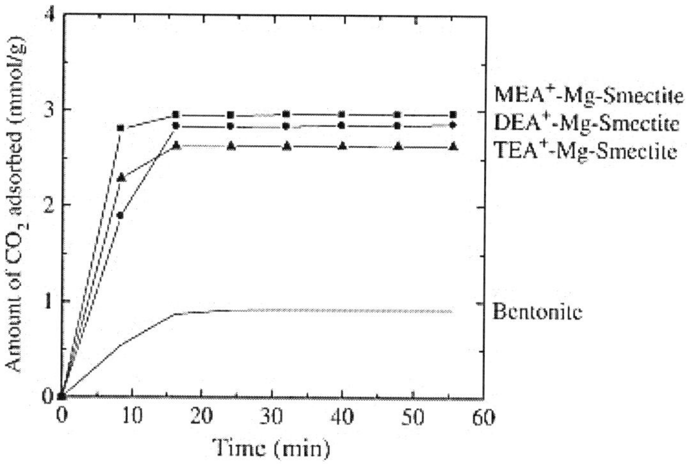

Figure 3. Effect of the molar mass of amines on rate of CO_2 uptake on untreated bentonite and bentonite clay adsorbents modified by amines at 298 K and 10 bar [73]. Copyright Elsevier B.V.

CONCLUSION

The current wave of interest in the intensification for search for ways of conserving and storing energy is not only aimed at finding improvements for energy efficiency but also geared towards a complete switch to clean and renewable energy. Many studies have been conducted worldwide in search for cost-effective materials for clean energy storage applications. Among these materials, clay and clay-supported materials have been in the forefront for clean energy storage applications. Apart from being abundant and environmentally benign, clay possess a variety of properties that are attractive for different kinds of energy storage applications. Depending on the type of energy storage technology, clay materials can either be used in their pristine form or as a support material. The discussion presented in this review has clearly shown that clays can effectively be applied in a range of energy storage technologies that span from thermal heat storage, storage of gases, uses in batteries and super capacitors as well as in fuel cell technology.

REFERENCES

[1] M. I. Abdou, A. M. Al-sabagh and M. M. *Egyptian Journal of Petroleum* 22, 53 (2013).
[2] J. Konta, *Appl. Clay Sci.* 10, 275 (1995).
[3] F. Bergaya and G. Lagaly, Handbook of Clay Science, *Elsevier*, Amsterdam (2013).
[4] C. T. Chiou and D. W. Rutherford, *Clays Clay Miner.* 45, 867 (1997).
[5] S. Bayülken, E. Başçetin, K. Güçlü and R. Apak, *Environ. Prog. Sustain. Energy,* 30, 70 (2010).
[6] O. M. Sadek and W. K. Mekhamer, *Thermochim Acta* 363, 47 (2000).
[7] O. M. Sadek and W. K. Mekhemer, *Thermochim Acta* 370, 57 (2001).
[8] D. Fernandes, F. Pitié, G. Cáceres and J. Baeyens, *Energy* 39, 246 (2012).
[9] J. Xu, R.Z. Wang, Y. Li, *Sol. Energy* 103, 610 (2014).

[10] K. Pielichowska and K. Pielichowski. *Prog. Mater Sci.* 65, 67 (2014).
[11] N. Sarier, E. Onder, S. Ozay and Y. Ozkilic, *Thermochim Acta*, 524, 39 (2011).
[12] S. A. Memon, T. Y. Lo, X. Shi, S. Barbhuiya and H. Cui, *Appl. Therm. Eng.* 59, 336 (2013).
[13] R. A. Shiggeishi, C. H. Langford and B. R. Hollebone, *Sol. Energy* 23, 489 (1979).
[14] R. Gopal, B. Hollebone, C. Longford and R. Shigeishi, *Sol. Energy* 27, 289 (1981).
[15] J. Jänchen, D. Ackermann, H. Stach, W. Brösicke, *Sol. Energy* 76, 339 (2004).
[16] W. K. Mekhamer, *Arabian J. Chem.* (2011), In press; DOI: 10.1016/j.arabjc. 2011.03.021.
[17] B. Givoni, *Sol. Energy* 19, 617 (1977).
[18] A.P. Sturman and H. A. McGowan, *J. Arid Environ.* 73, 74 (2009).
[19] A. Gabrielsson, M. Lehtmets, L. Moritz and U. Bergdahl, Swedish Geotechnical Institute, Report No. 53, Linköping (1997).
[20] M. Reuss, M. Beck and J. P. Müller *Sol. Energy,* 59, 247 (1997).
[21] D. J. Durbin and C. Malardier-Jugroot, *Int. J. Hydrogen Energy,* 38, 14595 (2013).
[22] A. W. C. van den Berg and C. O. Areán, *Chem. Commun.* 6, 668 (2008).
[23] A. Azzouz, *Int. J. Hydrogen Energy,* 37, 5032 (2012).
[24] F. Campos, L. de la Torre, M. Román, A. García and A. A. Elguézabal, *J. Ceram. Process.* Res. 9, 482 (2008).
[25] A. Gil, R. Trujillano, M. A. Vicente and S. A. Korili, *Int. J. Hydrogen Energy,* 34, 8611 (2009).
[26] A. Szücs, Z. Király, F. Berger and I. Dékány, *Colloids Surf., A,* 139, 109 (1998).
[27] J. Jin, C. Zheng, and H. Yang, *Funct. Mater. Letters,* 07, 1450027 (2014).
[28] M.M. Shaijumon, N. Bejoy and S. Ramaprabhu, *Appl. Surf. Sci.* 242, 192 (2005).
[29] M. A. Salam, Y. Lwin and S. Suriati, *Adv. Mater. Res.* 626, 173 (2013).
[30] M. A. Salam, S. Sufian and T. Murugesan, *Chem. Eng. Res. Des.* 91, 2639 (2013).
[31] C. Ruiz-García, R. Jiménez, J. Pérez-Carvajal, A. Berenguer-Murcia, M. Darder, P. Aranda, D. Cazorla-Amorós and E. Ruiz-Hitzky, *Science of Advanced Materials*, 6, 151 (2014).
[32] C. Ruiz-García, J. Pérez-Carvajal, A. Berenguer-Murcia, M. Darder, P. Aranda, D. Cazorla-Amorós and E. Ruiz-Hitzky, *Phys. Chem. Chem. Phys.* 15, 18635 (2013).
[33] H. B. Dai, X. D. Kang, and P. Wang, *Int. J. Hydrogen Energy,* 35, 10317 (2010).
[34] H. Tian, Q. Guo and D. Xu, *J. Power Sources*, 195, 2136 (2010).
[35] R. K. Stoessell and P.A. Byrne, *Clays Clay Miner.* 30, 67 (1982).
[36] K. Kosuge, *Journal of the Clay, Science Society of Japan*, 33, 215 (1994).
[37] D. J. K. Ross and R. M. Bustin, Mar. *Petrol. Geol.* 26, 916 (2009).
[38] D. Liu, P. Yuana, H. Liu, T. Li, D. Tan, W. Yuan and H. He, *Appl. Clay Sci.* 85, 25 (2013).
[39] J. B. Curtis, *AAPG Bulletin* 86, 1921 (2002).
[40] T. Zhang, G. S. Ellis, S. C. Ruppel, K. Milliken and R. Yang, *Org Geochem.* 47, 120 (2012).
[41] X. C. Lu, F. C. Li and A. T. Watson, *Fuel,* 74, 599 (1995).
[42] A. L. Cheng, and W. L. Huang, *Org Geochem.* 35, 413 (2004).
[43] M. M. Thackeray, C. Wolverton and E. D. Isaacs, *Energy and Environmental Science,* 5, 7854 (2012).
[44] E. Ruiz-Hitzky and P. Aranda, *Adv. Mater.* 2, 545 (1990).

[45] N. Shubha, R. Prasanth, H. H. Hoon and M. Srinivasan, *Mater. Res. Bull.* 48, 526 (2013).

[46] R. Prasanth, N. Shubha, H. H. Hng, and M. Srinivasan, *Eur. Polym. J.* 49, 307 (2013).

[47] M. Deka and A. Kumar, *J. Power Sources, 196*, 1358 (2011).

[48] M. J. Koh, H. Y. Hwang, D. J. Kim, H. J. Kim, Y. T. Hong and S. Y. Nam, *J. Mater. Sci. Technol.* 26, 633 (2010).

[49] Y. W. Chen-Yang, Y. T. Chen, H. C. Chen, W. T. Lin and C. H. Tsai, *Polymer,* 50, 2856 (2009 .

[50] P. Aranda, M. Darder, R. Fernández-Saavedra, M. López-Blanco and E. Ruiz-Hitzky, *Thin Solid Films*, 495, 104 (2006).

[51] A. Bakandritsos, T. H. Steriotis and D. Petridis, *Chem. Mater.* 16, 1551 (2004).

[52] L. Duclaux, E. Frackowiak, T. Gibinskp, R. Benoit and F. Begun, *Mol. Cryst. Liq. Cryst. Sci. Technol.* 340, 449 (2000).

[53] R. Fernández-Saavedra, P. Aranda, and E. Ruiz-Hitzky, *Advanced Functional Materials*, 14, 77-82 (2004).

[54] R. Fernández-Saavedra, P. Aranda, K. A. Carrado, G. Sandí, S. Seifert,and E. Ruiz-Hitzky, *Current Nanoscience,* 5, 506-513 (2009).

[55] M. S. Wang, L. Z. Fan, M. Huang, J. Li and X. Qu, *J. Power Sources,* 219, 29 (2012).

[56] A. Gómez-Avilés, M. Darder, P. Aranda, and E. Ruiz-Hitzky, *Appl. Clay Sci.* 47, 203 (2010).

[57] R. Gosalawit, S. Chirachanchai, S. Shishatskiy and S. P. Nunes, *J. Membr. Sci.* 323,337 (2008).

[58] C. Felice, S. Ye, and D. Qu, *Ind. Eng. Chem. Res.,* 49, 1514–1519 (2010).

[59] M. M. Hasani-Sadrabadi, E. Dashtimoghadam, F.S. Majedi, K. Kabiri, M. Solati-Hashjin and H. Moaddel, *J. Membr. Sci.* 365, 286 (2010).

[60] Z. Hu, G. He, S. Gu, Y. Liu, and X. Wu, *J. Appl. Polym. Sci.* (2014) DOI: 10.1002/APP.39852.

[61] L. Unnikrishnan, P. Madamana, S. Mohanty and S.K Nayak, *Polym. Plast. Technol. Eng.* 51, 568 (2012).

[62] I. Nicotera, A. Enotiadis, K. Angjeli, L. Coppola and D. Gournis, *Int. J. Hydrogen Energy,* 37, 6236 (2012).

[63] D. Manikandana, R. V. Mangalaraja, R. E. Avila, R. Siddheswaran, and S. Ananthakumar, *Mater. Sci. Eng., A,* 177, 614 (2012).

[64] Y. Pai and C. Tseng, *J. Power Sources*, 202, 28 (2012).

[65] R. M. G. Rajapakse, K. Murakami, H. M. N Bandara, R. M. M. Y. Rajapakse, *Electrochim Acta,* 55, 2490 (2010).

[66] S. S. Seo, *J. Appl. Electrochem.* 42, 539 (2012).

[67] M. Gupta, I. Coyle, K. Thambimuthu, CO2 capture technologies and opportunities in Canada. 1st Canadian CC&S Technology Roadmap Workshop, 18-19 Sept. 2003, Calgary, Alberta, Canada.

[68] IEA (2010) Energy Technology Perspectives 2010: Scenarios & Strategies to 2050. International Energy Agency (Paris, France) pp. 710.

[69] C. Volzone, J. O. Rinaldi and J. Ortiga, *Materials Research,* 5, 475-479 (2002).

[70] C. Volzone and J. Ortiga, *Process Saf. Environ. Prot.* 82, 170 (2004).

[71] C. Volzone and J. Ortiga, *J. Mater. Sci.* 35, 5291 (2000).

[72] A. Azzouz, E. Assaad, A. V. Ursu, T. Sajin, D. Nistor and R. Roy, *Appl. Clay Sci.* 48, 133 (2010).

[73] A. E. I. Elkhalifah, S. Maitra, M. A. Bustam, and T. Murugesan, *Applied Clay Science,* 83–84, 391–398 (2013).

In: Bentonite
Editor: Ajay Kumar Mishra
ISBN: 978-1-63482-142-1

Chapter 2

TRADITIONAL AND MODERN USES OF BENTONITE CLAY AND IMPLICATIONS ON THE ENVIRONMENT

Derrick S. Dlamini*
Department of Chemistry, University of Swaziland, Swaziland

ABSTRACT

Bentonite clay has outstanding physico-chemical properties which make it useful in a wide range of areas of application such as water treatment, medical, groundwater barrier, farming, and construction to mention just a few. This clay is known to be one of the most versatile and inexpensive treatments available. Some of the many health benefits of bentonite clay are: balancing bacteria in the digestive tract and improving nutrient assimilation. In this chapter, traditional and modern uses of bentonite clay as well as future potential applications of the clay are highlighted.

INTRODUCTION

There are various definitions of bentonite that have been given in the literature, but generally they all state that bentonite is a rock term used to designate a naturally occurring, very fine grained material largely composed of the clay mineral, montmorillonite. Figure 1 shows X-ray Powder Diffraction pattern of a raw sample which shows the presence of minerals montmorillonite, quartz and calcite.

The largest known and most active deposits come from Wyoming and Montana [1]. According to Hosterman et al. [2] bentonite includes any natural material dominantly composed of clay minerals in the smectite group. Bentonite from Wyoming consists of hydrous silicate of alumina, commonly known to as montmorillonite clay or sodium bentonite.

* Email address: derricksdlamini@gmail.com

Accordingly, a description and discussion of montmorillonite is applicable to bentonite as well. Bentonite, in addition to montmorillonite, contains a small portion of other mineral matter, usually quartz, feldspar, volcanic glass, organic matter, gypsum, or pyrite.

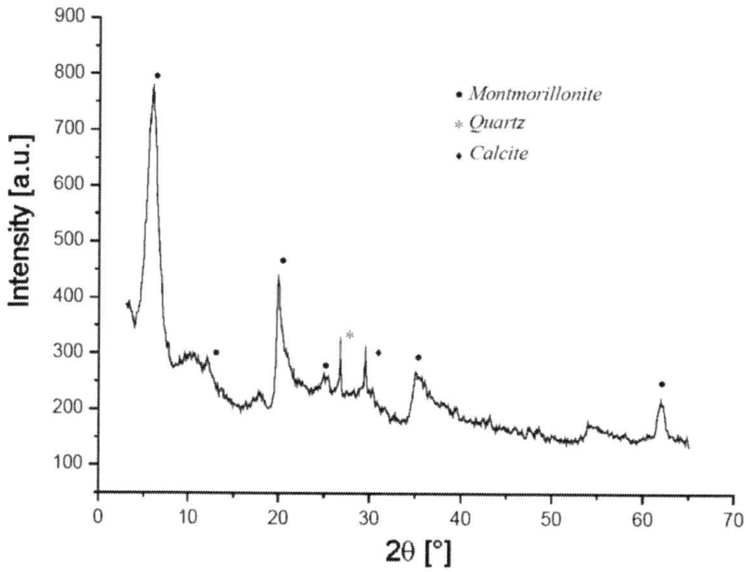

Figure 1. XRPD pattern showing the components of bentonite.

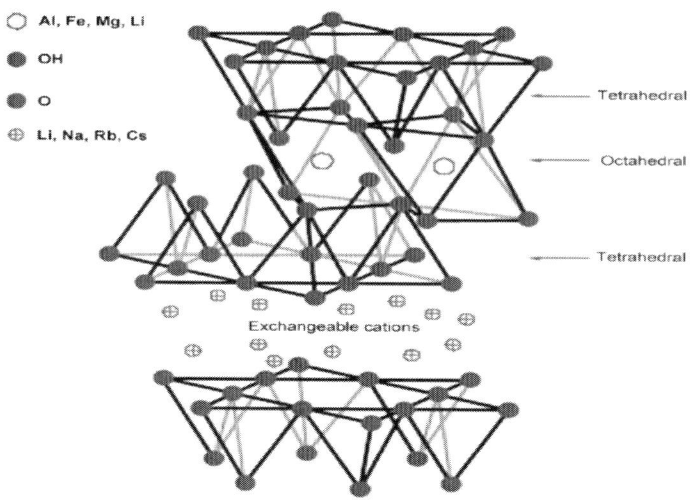

Figure 2. Structure of layered-silicates.

Chemically, montmorillonite is described as a hydrous aluminum silicate containing small amounts of alkali and alkaline earth metals. Structurally, montmorillonite is made of two basic building blocks, the aluminum octahedral sheet and the silica tectrahedral sheet as seen in Figure 1.

Bentonite is popular for its high-swelling properties; it can swell up to 16 times its original size and absorb up to 10 times its own weight when exposed to water [3]. Virta [3] also reported that sodium bentonite is used in kitty litter (25%), drilling mud (20%), binder in foundry molds (19%), iron ore pelletizing (13%), and other applications (23%). Calcium bentonite, a low or non-swelling variety, is relatively unimportant in Wyoming production. Bentonite, originally known as 'mineral soap' or 'soap clay', was named 'bentonite' in 1898 by Wilbur C. Knight for deposits in the Benton Shale near Rock River, Wyoming.

Calcium-Bentonite, also known as non-activated bentonite, is a type that is predominantly occupied by Ca^{2+} or Mg^{2+} ions in the intermediate layers. During activation (alkali-activation), the Ca^{2+} -ions in the intermediate layer are replaced with Na^{+}-ions. Sodium-Bentonite does however occur naturally, and is predominantly occupied by Na^{+}-ions in the gallery spacing. In addition to Na^{+}-ions, Ca^{2+} or Mg^{2+} -ions can also occur generally in Na-bentonite in varying concentrations. In Organophilic Bentonite or organoclays, the cations in the intermediate layers are replaced by polar organic molecules known as modifiers. Depending on the nature of the modifier, they may be hydrophobic and can swell in organic solvents. Organoclays are discussed in detail in section 3.2.8.

Industrially Attractive Properties of Bentonite

A great number of bentonite uses in industry results from its swelling characteristic when in contact with water. The water–bentonite combinations can perform a lot of jobs in industry because physical properties of the mixtures change as the water-to-clay ratio changes. The swelling capacity of the clay is a characteristic of great importance and is controlled mostly by the types of associated ion exchangers.

When a fairly small amount of water is mixed with a huge amount of bentonite, the resultant mixture forms a sticky mass which has adhesive properties. This physical property results from insufficient water content to fulfill the oriented water required by the clay. The plastic properties of a clay-water binary system increase as the ratio of water to clay increases. This is because when each montmorillonite flake has adsorbed its maximum amount of water, extra water molecules in the system act as lubricants for the hydrated particles. It takes a long period of time for the clay to develop a completely hydrated state, even if the necessary water is available, since the clay is made up of 'impermeable' layers which enclose the exchangeable ions. The hydration process can be hastened by agitation or stirring of the suspension. In this process, the clay flakes are and this substantially enhances the clay swelling.

Another industrially important property of bentonite is its reactivity. Unlike many other materials that have a large surface area, the surfaces of bentonite are principally populated by oxygen atoms which promote a high degree of reactivity with other compounds which makes bentonite to be an excellent absorbent. A well dispersed bentonite in water has a high potential to carry on other reactions.

Uses of Bentonite Clay

Suspensions of bentonite in water are broadly used in industry. Its known cha high plasticityhigh water affinity and it also has high shrinkage and swelling characteristics. The viscosity of an aqueous medium is augmented significantly when bentonite is added and mixed in amounts generally less than 10% (w/w). These suspensions contain bentonite particles which have been dispersed into the colloidal particle range, and therefore, they have similar properties and respond like other colloidal suspensions. Some of the many applications of bentonite suspensions include; drilling mud for rotary drills, fire retardant gel, and media for suspending materials which range from medicines that are taken internally to lumps of coal as part of a float sink or separatory process. The uses of bentonite spread into more modern applications like paper-making technology, cut-off walls, wastewater purification and has recently spread into different aspects of nano-technology. The application of bentonite suspension is so broad because theviscosities of the suspensions best suited for any application are easily controlled by adjusting the bentonite–water ratio. Some of the common uses of bentonite are discussed below.

Medicinal Uses

Below is evidence that provides sufficient information about the usefulness and application of bentonite in practical fields of medicine. Bentonite is used internally and externally for various health issues.

Treatment and Prevention

Bentonite can be used both for treatment and prevention. It is well known for treating gum diseases, ulcers, rashes, dysentery, hemorrhoids, infected wounds and bites. For slow healing of ulcerated wounds, abrasions, crushed fingers and cuts, the entire wound area is covered thinly with hydrated Calcium bentonite clay. Calcium bentonite clay is actually referred to as the healing clay. For a laceration or a cut (minor and severe), dry powdered healing clay is poured into the wound to rapidly stop bleeding and prevent infection. Dr Meyer Camberg used clay to neutralize arsenic poisoning.

During the First World War, the Russian soldiers received 200 grams of clay along with their rations, and it was added to mustard in several of the French regiments to prevent dysentery which reportedly ravaged nearby regiments. Furthermore, Russian scientists used clay to protect their bodies from radiation when working with nuclear material. Bentonite clay was the material chosen to dump into the reactors after the nuclear meltdown.

The healing clay works by two mechanisms: adsorption and absorption. Adsorption is defined as a process by which charged particles of other substances combine with charged particles on the outer surface of the clay molecule. Clay molecules carry a negative electrical charge while toxins, e.g., most bacteria, viruses, fungi, , toxic chemicals,etc usually carry a positive charge. When the clay is taken into the human body, the positively charged toxins are attracted to the negatively charged surfaces of the clay molecule in a chemical adsorption mechanism. An exchange reaction occurs in which the clay mineral ions are swapped for the ions of the toxic substance, a process called ion exchange. When the ion exchange has been completed, the clay will hold onto the bio toxin until it is excreted by the body.. Adsorption is generally a fast process under desirable conditions.

Absorption, on the other hand, is a slower and more complex process compared to adsorption. In this case the clay molecule acts like a sponge and draws other substances into its internal structure. Absorption can only occur when the foreign substance has undergone a chemical change and is then allowed to enter the molecular inner structure of the clay. Once the foreign substance has undergone the needed chemical change, it enters into the spaces between the clay's inner structures. So the toxins that were previously only sticking to the surface of the clay's outer structure through ionic bonding are now pulled inside the clay molecule. This is one of the reasons why absorptive clays are labeled as mobile layered or expandable clays. The more substances absorb into the inner structure of the clay, the more the clay expand and its layers swell.

Calcium bentonite clay is a powerful antioxidant. It is rich in diastases enzymes, which is responsible for its ability to fix free oxygen and purify and enrich the blood. Clay can be taken to balance body pH in humans. A healthy body maintains a neutral pH at around 7.4. When the human body is acidic (pH less than 7) diseases are more likely to flourish. Most people have an acidic pH and calcium bentonite clay has an alkaline pH of 9.7, hence it can be taken to balance their pH. Bentonite clay has many healing capabilities due to its alkaline pH and high negative ionic charge.

Calcium bentonite clay can relieve a congested and overworked liver. Weight gain and depression are just two common symptoms of a congested and overworked liver. When the liver cannot remove toxic substance it gets congested it will remain that way and get worse until it gets detoxified and re-energized. Bentonite clay can cleanse and rebuild the liver. General symptoms of a congested, overworked liver are listed below:

- Stress/anxiety
- Elevated cholesterol
- Skin conditions
- Skin irritation
- Depression
- Sleep difficulties
- Indigestion
- Kidney damage
- Heart damage
- Brain damage
- Hypothyroidism
- Chronic fatigue
- Weight gain
- Poor memory
- Mental fog
- Blood sugar disorders
- Allergies
- Obesity
- Etc.

Bentonite clay supports eradication of allergies. It is known that allergies are triggered by the release of histamines. This happens when the body cannot produce sufficient antihistamines to fight allergic reactions when the liver is overworked by toxins. In other applications, plain bentonite can be used to make baby powder. The baby podwer is used to soothe the skin when there is infection or redness. It can help speed recovery in this area if prapared into a clay "mask". In addition, bentonite clayis also used to treat various skin conditions which include, among others, eczema, acne, cellulite, burns, insect bites, and fine wrinkles. Bentonite clay can even balance bacteria in the digestive tract. It can strengthen the immune system. It does that by improving digestion; as a result a healthy working digestive tract absorbs nutrients more efficiently. Bentonite clay helps create a better environment in the body so it breaks down fat, nutrients, and sugar more effectively and that results in weight loss.

Detoxification

Upon contact with fluid, bentonite electrical components change, giving it the power to absorb toxins. The electrical charge is credited for its ability to absorb and remove toxins, heavy metals, impurities, and chemicals and rid the body of the harmful effects of exposure to radiation. Using clay internally and externally can assist in the removal of radioactive isotopes from the body. Therefore, bentonite clay is useful in the removal of toxins from our body system.

The clay particles naturally work to absorb heavy metals and other toxins in the gut thus improving intestinal health and preventing the toxins from being absorbed into the body. It has been reported that it is useful to add fiber such as Psyllium husks (when using clay internally) to help move the clay along through the digestive tract [4]. Clay baths, poultices and topical applications can also help the human body detox through the skin pores, credit to clay's strong drawing powers. Bentonite clay can draw out gangrene, heavy metals and toxins [5] through a pulsing action. The composition of bentonite naturally replenishes the body's nutritional needs while removing toxins [5].

The very small particles of bentonite gives a large surface area in proportion to the volume used, thereby enabling it to pick up many positively charged particles (toxins). To obtain maximum effectiveness in the human body, bentonite must be put in a liquid colloidal gel form. In such a state, it swells and increases its surface area. Ditter et al., [6] studied the capability of various adsorbents to bind with toxins and bentonites reportedly showed the best results, followed by kaolin-pectin (Kaopectate). Calcium bentonite is one of the safest, easiest, and cheap detoxing substances available. It is a non-invasive treatment, and when properly used, it causes no harm and hence rapid healing and reduction in pain, irritation, and inflammation often results.

Industrial Uses

Bentonite-water mixtures are used for bonding, plasticizing and suspending. This versatility is made possible chiefly by adjusting the quantities of the two components. A bonding agent must have certain adhesive properties, and these are developed when bentonite is mixed with only a small amount of water [7]. When oriented water is shared by adjacent clay flakes, and when this mixture is added as a constituent to a combination of materials, rigidity rather than plasticity is developed in the final product. For a bonding agent, generally both water and clay are mixed into the finished product in amounts less than 5 % (w/w) The swelling potential produces a tight contact with confining structures, which makes bentonite ideal as a pack of rock-deposited radioactive canisters and for borehole and shaft sealing. Even at high densities between 1.1 and 1.3 g/cm^3, the hydraulic conductivity of smectite gels is still sufficiently low to make them useful as effective grouts for sealing rock fractures [7].

Binder

All natural forms of bentonite are used by foundries; the strengths of the sand mixtures, whether green, hot, or dry, can be adjusted by varying the mixtures of the clay types. The bonding properties that are ideal for use with insulation materials are generally comparable to those in foundry sands. Each industry has strength requirements for its products at various stages of processing, and in each case the raw materials consist of a large number of particles which need to be bound together. The bentonite that binds these products well is one which disaggregates into extremely small particles to provide a large surface area with many available binding sites. This allows maximum contact with other product components and that enhances binding. The taeonite industry provides an interesting example of the use of bentonite as a binding agent. One example of the use of bentonite as a binding material is in the processing of low-grade iron ore in an economic operation. In this process, the ore is crushed into a fine powder and then the iron particles are separated from other rock particles by using a magnetic field process. Due to the difficulty in handling fine powder, the powdered and up-graded iron ore is pelletized to aid further processing and excellent results are obtained when bentonite is used to bond the pellets.

Drilling Mud

One of the largest and best known uses of bentonite suspensions is in drilling mud. The functions of drilling mud include the following; removal of cuttings from the bore hole, prevention of blow-outs, strengthening of the well bore and reduction of fluid loss from the mud, prevention of weighting agents and cuttings from settling down the well when circulation stops, and lubrication and cooling of the drilling bit and string to mention just a few. These favorable properties in the mud depend on the viscosity and gel characteristics that are contributed by the bentonite. Impermeability of the filter cake is due in part to the fact that the adsorbed water on bentonite is very tightly bonded because bentonite is hydrophilic and also because the many small clay flakes act together to form a nearly perfect lining of the bore hole.

Fire Retardant Material

Another application of bentonite suspensions is in fire retardant material, predominantly in combating forest fires. When a stream of water is directed at vegetation, such as trees, much of the water ricochets off and is lost owing to its high fluidity. When the viscosity of the water stream is increased with bentonite, a greater proportion of liquid is retained on the vegetation. Because of the thicker viscosity, suspensions can be dropped from helicopters to aid fire-fighting especially in forests. The advantages obviously are greater coverage within a shorter period of time, and accessibility to virtually all locations.

Control of Water Flow

Due to bentonite's good water-adsorption capacity, it is used as a material to impede the flow of water [7]. The flow of water clay can be controlled the by two methods. In the first case, when dry hydrophilic clay comes in contact with a source of water, the water becomes adsorbed by the clay seizes to flow. There is however a limit to the amount of water which can be absorbed by a given amount of clay. This called a point of saturation. This is an advantage because saturation results into a second water-impedance mechanism. Once the clay is hydrated or swollen to a saturation point, it becomes impermeable to the flow of any additional water. In this way it is likely for small amount of clay to stop or impede a large amount of water if the clay is applied advantageously. Sodium smectite clay, a commercially available form of bentonite powder, serves as a very effective water flow barrier [7].

Carrier Materials

Bentonite is frequently used in certain industrial problems that can be solved best by the use of materials that have a large surface area. An example of such applications is the use of bentonite as a carrier material for chemicals, such as insecticides. Using carrier materials with a large surface area like bentonite has several advantages which include the distribution of insecticides when dispersed over a large surface area. Other advantages include increasing the resident time of the pesticide on the plant, and also increases the effectiveness of the pesticide by keeping it in the most advantageous position; that is, on the target (vegetation) rather than on the ground.

Emulsion Stabilizer

It is highly necessary for emulsion stabilizers to possess a large surface area. An example in which the large surface area supplied by bentonite becomes useful is when it is applied in industry as an emulsion stabilizer. Emulsions are formed by agitation and blending of two immiscible liquids, such as oil and water. One liquid (e.g., water), becomes a continuous phase and the other liquid (e.g., oil) forms as droplets or globules within it. The role of each liquid in the emulsion is determined by the ratio in which the two liquids are mixed. The stability of emulsion is increased when the oil-water interface is strengthened. Bentonite has a significant stabilizing effect on emulsions when it is added to the mixture in stages preceding agitation. This is so because the clay flakes form a thin film at the oil-water interface, and the large surface area of the clay helps in the entrapment of the oil globules as they are broken down by mechanical agitation.

Reactions

So far this chapter has emphasized on the ability of bentonite to react with water. There are industrial applications of bentonite which rely upon its ability to react with other materials. Although reactions do occur in the uses discussed thus far, they are regarded as secondary reactions to the application. There are applications where primary reactions take place. They are primary to the bentonite application, whereas reactions between clay and other materials may be considered as secondary in some of the applications discussed earlier. For example, when bentonite is used as an insecticide carrier, there are probably reactions which occur in addition to the "plating-out" of the insecticide on the clay mineral surfaces. Reactions between bentonite and amines were initially studied by Hauser and Leggett [8] and later by Jordan [9]. These investigations found that bentonite surfaces could be entirely coated with organic material by the substitution of inorganic cations by the amine. This was accomplished on less hydrophilic clay that have lost its ability to swell in water. The clay-organic product swell in a number of organic solvents, and because of this property, a market in industry was created for the products. They are used as gelling agents in paints, greases and lubricants, and in other oil-based media.

Creation of Organophilic Clay

Organophilic clay, commonly called nanoclay, can be formed by exchanging organic cations such as alkyl ammonium or phosphonium ions into the clay interlayer [10]. The high cation exchange capacity of sodium bentonite makes it a suitable platform for surface modification. The naturally occurring Na^+ ion is substituted with an organic cation, which can be typically introduced during different processes. Several types of organic chemicals have been utilized for the modification of sodium bentonite clay. The onium salt replaces the metal cations located between the layers and increases the interlayer or basal spacing [11] as shown in Figure 3.

The positive charge of the amines allows for an effective bond to the negatively charged clay. The resulting product is hydrophobic and thus will not swell in water but has a high affinity for low soluble organic matter. There are numerous variations of these products that allow for the design of a media that will be more selective on certain organics based on their solubility.

Polymer–clay nanocomposites are widely used in industry. The hydrophilic nature of the clay surfaces impedes their homogeneous dispersion in the organic polymer phase [12] during synthesis. However, homogeneous dispersion can be improved if the clay used is transformed into organoclay. That is another importance of organoclay in industry.

Organoclay can be used as adsorbents for organic matter such as oil, grease and other high molecular weight hydrocarbons from water, and it is preferable over granular activated carbon which adsorbs organic matter through the porous surface. This mechanism requires right size of the adsorbed organic matter to squeeze through the pores. On the other hand, organoclay utilize the partition media created during the surface modification process. It is noteworthy to highlight that the organic matter removal efficiency of organoclay is directly proportional to the solubility (in water) of the organic contaminants present. Initial applications focused on inland and offshore drilling operations traditionally utilized granular activated carbon to treat waste waters before discharge.

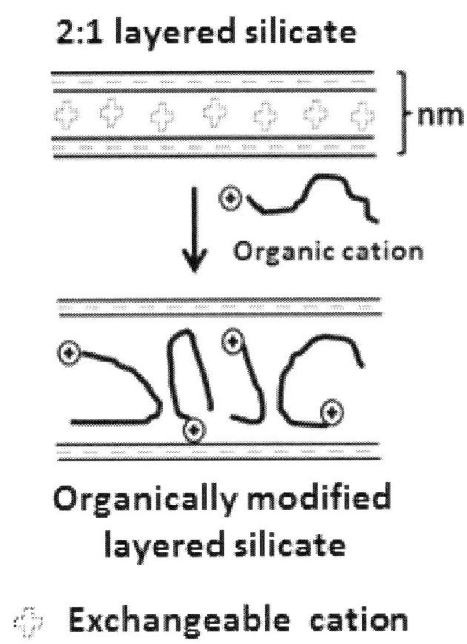

Figure 3. Schematic of clay modification with an organic cation.

Although granular activated carbon has proven to be a very effective media for adsorbing organic matter, unfortunately it is subject to blinding and premature exhaustion in the presence of petroleum based organic compounds and other high molecular weight water contaminants.

Clarifying Agent

Bentonite has an application in the brewing industry. It is known to have beneficial properties as clarifying agents for products such as wines and beers. Many of the impurities, or sediment, present in these liquids during their manufacture are extracted using bentonite. Calcium bentonite has also been used for many years to decolorize oils. It has been reported that calcium is better than sodium bentonite for this use but the exact reasons for this are not fully understood [8, 13]. The decolorizing process is fairly simple. Some oils are clarified by allowing percolation through beds of bentonite in which selective adsorption mechanism of some organic types onto the clay mineral surfaces occurs [9].

Agricultural Uses

Over a decade ago, farmers in the Czech Republic began to use non-conventional feed supplements such as peat to prevent economic losses caused by diarrhea in piglets [14, 15]. The adsorbent capability of peat alleviates the adverse effects of toxins present in feeds, and thus protects highly sensitive mucosa of the digestive tract, stomach and intestines, particularly in the young farm animals like piglets [16 – 21]. Fuchs et al., [22] observed that supplementation of peat with minerals can improve the health status in piglets. It is the adsorbent capability of the clay mineral that initiates detoxification and prevention of diarrheal diseases in animals. Therefore, the bentonite can be used as an ingredient to create pellets of animal feed from coarse ground components. As reported by Southern et al., [23], bentonite supplementation (5%) of the diet given to chickens which are fed nutrient deficient diets can enhance the feed intake by the chickens, and the weight gain which may be decreased by these deficiencies.

Impurities

All natural occurring bentonites have small fragments of impurities (feldspar, calcium carbonate, gypsum, quartz, etc.). These fragments are termed“grit”, and may constitute about 5-10% of prepared commercial bentonite [24].

The bentonite we use, known as Sodium Bentonite, is a high purity, airfloated bentonite consisting of micronized particles. Bentonite is one of the volcanic ashes; it is a product of the earth which contains a small amount of aluminum as well as other trace minerals. It also contains Hydrous Aluminum (Al_2O_3) Silicate that is actually part of the bentonite crystal which is not leachable by any means. The aluminum element is central to the bentonite molecule holding the negative charge around it as a cloud of oxygen molecules.

In effect, part of the negative charge of bentonite emanates from the aluminum. Removal of the aluminum from the molecule would result in a loss of the drawing properties of the molecule. Fortunately, the human bodies do not have sufficient digestive juices to split apart such a highly negatively charged molecule.

Bentonite Clay Effects on Health Problems and the Environment

Five workers who were exposed to bentonite or Fuller's earth [25 – 28] for periods ranging from less than 15 years to more than 40 years developed dyspnoea. In some cases this was accompanied by cough and sputum, and pneumoconiosis. The general characteristics of the pneumoconiosis of the workers, as revealed on autopsy, were reported to be as follows:

- Emphysema
- Lesions
- Numerous firm black patches
- Nodules 1–20 mm in diameter
- Dilated air sacs
- Adhesions
- Isolated cavities filled with "black sludge"
- and other evidence of physical deterioration

Three of the studies performed by Sakula [27] and Gattner [29] reported radiological evidence of significant damage to lungs following exposures to bentonite or montmorillonite dust of 10-42 years although no autopsy was presented to support the evidence. Gattner [29] also reported radiological proof of possible slight damage to lungs in three workers following 1 to 3 years of exposure. One of the important components of all clays is crystalline silica, the amount of which varies greatly. Crystalline silica causes various lung diseases, including silicosis and lung cancer hence it is fair to say its content is a decisive factor in clay-induced adverse health effects. To the best knowledge of the author, there are too little or no studies in humans of the effects of long-term oral exposure to clay. To add to that, no adequate studies on the carcinogenicity of bentonite are available.

Since bentonite clay has high absorptive properties (as already discussed), it has the potential to absorb vitamins and other (positively charged) supplements in the digestive tract, preventing them from finding their way into the blood stream. It is therefore recommended to take supplements either 2 hours before, or 2 hours after the ingestion of bentonite clay. Bentonite clay may also inhibit the efficacy of prescription medications, and so it should be taken at a different time of the day than the clay gel.

Bentonite clay has low toxicity to aquatic species. There is no reason to believe that the mining or processing of bentonite poses significant toxicological dangers to the environment. However, there are environmental issues just like in any other large-scale mining operation: physical disturbance to the land, excessive stream sedimentation, and similar destructive processes resulting from the large-scale mining and processing of clays.

CONCLUSION

Calcium bentonite clay is a great absorbent and will absorb positively charged toxins and impurities and ignore negatively charged nutrients. It is therefore a detox agent which can be applied externally (topically) or dissolved in water and taken internally. It eliminates unhealthy positively charged toxins, most bacteria, and other unhealthy noxious substances. In summary, this is the best, purest, most effective natural clay available to date. Many reports of bentonite use in animal feed have been documented. Unfortunately the world's medical community does not promote it, but that is because there hasn't been extensive tests done to prove the validity of the many negative claims made for clay.

REFERENCES

[1] R. E. Grim, "Clay Mineralogy." McGraw-Hill Book Company, INC, New York. 220-230; 361-364, (1953).

[2] J. W. Hosterman and S.H. Patterson, "Bentonite and fuller's earth resources of the United States." U.S. *Geological Survey Professional Paper,* 1522, 45, (1992)

[3] R.L. Virta, "Clays" United States Geological Survey, Mineral Commodity Summaries, United States Department of the Interior, 48–49.

[4] G. Steven, "Wyoming bentonite-a niche in mineral globalization." In Proceedings of the 3rd North American Industrial Minerals Annual Meeting, Atlanta, pp. 20, October 28-29, (1998).

[5] M. A. Asad, S. Kar, M. Ahmeduzzaman and M.R. Hassan, "Suitability of bentonite clay: an analytical approach." *International Journal of Earth Science,* 2(3), 88-95, (2013).

[6] B. Ditter, R. Urbaschek, and B. Urbaschek, "Ability of various adsorbents to bind endotoxins in vitro and to prevent orally induced endotoxemia in mice." *Gastroenterology,* 84, 1547–1552, (1983).

[7] R. Pusch, "Use of bentonite for isolation of radioactive waste products." *Clay Minerals,* 27, 353-361, (1992).

[8] E.A. Hauser and M.B. Leggett, "Color reactions between clays and amines." *J. Amer. Chem. Soc.,* v. 62, pp. 1811-1814, (1940).

[9] J. W. Jordan, "Organophilic bentonites, I, Swelling in organic liquids." *J. Phys and Colloid Chem.,* 53, 294-306, (1949).

[10] R. M. G. Rajapakse, K. Kurakami, H.M.N. Bandara, R.M.M.Y. Rajapakse, K. Velauthamurti, and S. Wijeratne, *Electrochim. Acta,* 55(7), 2490–2497 (2010).

[11] D.S. Dlamini, S.B. Mishra, A.K. Mishra, and B.B. Mamba "Ethylene Vinyl Acetate and Polycaprolactone–Organoclay Nanocomposite: Thermal, Mechanical and Morphological Properties." *J. Inorg. Organomet Polym.,* 21, 229–236, (2011).

[12] Ladhari, H.B. Daly, H. Belhadjsalah, K.C. Cole and *J. Denault, Polym. Degrad. Stab.,* 95, 1–11 (2010).

[13] F. Chisholm, "Bentonite in industry." *Mines Magazine,* July, pp. 30-42, (1960).

[14] L. Matlova, L. Dvorska, J. Bart, J. Bartos, W.Y. Ayele, M. Alexa, and I. Pavlik, "Mycobacteria isolated from the environment of pig farms in the Czech Republic during the years 1996 to 2002." *Veterinarni Medicina,* 48: 343–357, (2003).

[15] Pavlik, L. Matlova, L. Dvorska, J. Bart, L. Oktabcova, J. Doceka, and I. Parmova, "Tuberculous lesions in pigs in the Czech Republic during 1990–1999: occurrence, causal factors and economic loses." *Veterinarni Medicina,* 48, 113–125, (2003).

[16] M. Kuhnert, K.P. Bartels, S. Kroll, and N. Lange, "Veterinary pharmaceuticals containing humic-acid fortherapy and prophylaxis for gastrointestinal-diseases of dog and cat." *Monatsh. Vet. Med.,* 46, 4–8, (1991).

[17] T. D. Lotosh, "Experimental bases and prospects forthe use of humic acid preparations from peat in medicine and agricultural production." Nauchnye Doki Vyss. *Shkoly Biol. Nauki,* 10, 99–103, (1991).

[18] L. M. Stepchenko, L.V. Zhorina, and L.V. Kravtsova, "The effect of sodium humate on metabolism and resistance in highly productive poultry (in Russian)." Nauchnye Doki Vyss. *Shkoly Biol. Nauki,* 10, 90–95, (1991).

[19] Herzig, J. Hamp, H. Docekalova, B. Pisarikova, and J. Vlcek. "The effect of sodium humate on cadmium deposition in the chicken organs." *Veterinarni Medicina,* 39, 175–185, (1994)

[20] A. M. Beer, J. Lukanov, and P. Sagorchev.: The influence of fulvic and ulmic acids from peat, on the spontaneous contractile activity of smooth muscles. *Phytomedicine,* 7, 407–415, (2000).

[21] P. I. Zagorchev, A.M. Beer, and J.B. Lukanov. Effect of fractions of water extract from peat with a specific molecularweight on the spontaneous contractile activity of guinea pig stomach smooth muscles. *Folia Med.(Plovdiv),* 42, 52–56, (2000).

[22] B. Fuchs, J. Orda, J. Pres, and M. Muchowicz, "The effect of feeding piglets up to the 100th-day of their life with peat preparation on their growth and physiological and biochemical indices." *Arch. Vet. Pol.,* 35, 97–107, (1995).

[23] L. L. Southern, T.L. Ward, T.D. Bidner, and L.G. Hebert, "Effect of sodium bentonite or hydrated sodium calcium aluminosilicate on growth performance and tibia mineral concentrations in broiler chicks fed nutrient deficient diets." *Poultry Sci.,* 73, 848–854, (1994).

[24] S. Johnstone, and M. Johnstone, Minerals for the chemical and allied industries. Chapman and Hall. Second edition. London. 69-75, (1961).

[25] A. H. Campbell, and S.R. Gloyne, "A case of pneumoconiosis due to the inhalation of Fuller's earth." *J. Pathol. Bacteriol.,* 54: 75–79, (1942).

[26] H. O. Tonning, "Pneumoconiosis from Fuller's earth." *J. Ind. Hyg. Toxicol.,* 31: 41–45, (1949).

[27] Sakula, "Pneumoconiosis due to Fuller's earth." *Thorax,* 16: 176–179, (1961).

[28] A. R. Gibbs, and F.D. Pooley, "Fuller's earth (montmorillonite) pneumoconiosis." *Occup. Environ. Med.,* 51: 644–661, (1994).

[29] H. Gattner, "Die Bleicherde-Lunge." *Arch. Gewerbepathol Gewerbehyg,* 13: 508–516, (1955).

In: Bentonite
Editor: Ajay Kumar Mishra
ISBN: 978-1-63482-142-1

Chapter 3

ORGANO-BENTONITE ADSORBENTS FOR DYES AND OTHER ORGANIC POLLUTANTS IN WATER

Susmita Sen Gupta[1] and Krishna G. Bhattacharyya[2,*]
[1]Department of Chemistry, B N College, Dhubri, Assam, India
[2]Department of Chemistry, Gauhati University, Guwahati, Assam, India

ABSTRACT

Natural bentonite is abundantly available in many places and has gained importance as a low-cost adsorbent material for water purification. The adsorbing capacity of bentonite and its selectivity towards various pollutants could be modified substantially through the introduction of inorganic and organic substances into the clay structure. The hydrophilic character of the bentonite surface generally makes it less attractive as an adsorbent for non-ionic, non-polar organic pollutants in water. However, this problem has been sought to be tackled by docking large organic cations on bentonite surface through ion-exchange reactions and the resultant organo-bentonite has the potential to be very effective in adsorbing organic micro-pollutants from water such as various dyes, phenols and substituted phenols, etc. In this review, preparation and characterization of the organo-bentonites and their use as sorbents for dyes and other organic pollutants in water are dealt with based on the published work during the last ten years.

INTRODUCTION

Clays and Clay Minerals

Clays and clay minerals are found at or near the surface of the Earth. Clays are the mixtures of fine-grained clay minerals and clay-sized crystals like quartz, carbonate, and metal oxides. However, the term 'clay minerals' refer to phyllosilicate minerals and to minerals which impart plasticity to clay and which harden upon drying or firing (Guggenheim and Martin, 1995). Phyllosilicates are a large family of minerals that commonly show layered

[*] Corresponding author: E-mail: kgbhattacharyya@gmail.com

structures and include kaolin, smectite, chlorite, mica and serpentine groups. The particles of clay minerals may be crystalline or amorphous, platy or fibrous and may vary from colloid dimensions to those above the limit of resolution of an ordinary microscope (Deer et al., 1985).

Role of Clays in the Environment

Clays play a significant role in the environment by acting as a natural scavenger of pollutants by taking them either through ion exchange or adsorption or both as water penetrates downward through soil. In the adsorption process, adsorbate ions are bound to the clay surface by sharing one or several ligands (generally oxygen) with adsorbent cations as isolated complexes.

The adsorptive properties of clay minerals are thought to be due to their porous structure, large surface area and other properties, like cation exchange capacity. The edges and the faces of clays can adsorb different types of cations, anions as well as non-ionic and polar organic contaminants from water. Both Brönsted and Lewis type of acidity in clays (Tanabe, 1981) enhance the adsorption capacity of clay minerals to a great extent. It has been reported that if the uptake of pollutant is below CEC of the clays, the mechanism of adsorption is dominated by ion exchange, but the cause of adsorption becomes hydrophobic bonding if the uptake is more than CEC (Jamirah and Dental, 1993). The interactions between clays and the adsorbed pollutants have some stereo specificity as the clay surface imposes certain steric constraints on the adsorbed molecules due to their positions in the interlayer space. In general, the adsorption capacity depends on some primary factors, namely pH and temperature of the medium, nature of the clay minerals (adsorbent), presence of organic matter, nature of the adsorbates, etc. (Sheng and Boyd, 2000). The polarity and hydrophobic character of the surface of some bentonite clays enhance retention of organic pollutants (Mortland et al., 1986). Crains-Smith and Hartman (1986) reported that the crystal imperfections in clay minerals affect the adsorption and reaction processes on their surfaces.

The possible active sites in clays are (van Olphan, 1977): (i) Brönsted acid sites or proton donor sites, created by interactions of adsorbed or interlayer water molecules, (ii) Lewis acid sites or electron acceptor sites, occurring in absence of adsorbed water molecules, (iii) Oxidizing sites, arising from the presence of some cations (e.g., Fe^{3+}) in octahedral positions or due to adsorbed oxygen on surfaces, (iv) Reducing sites, occurring due to the presence of some cations (e.g., Fe^{2+}) and (v) Surface hydroxyl group, mostly found in the edges, bound to Si, Al or other octahedral ions.

Structural Features of Montmorillonite

Clay minerals are characterized by a continuous sheet structure made of layers of Si-O (± Al) tetrahedra. Each tetrahedron consists of a cation, T, coordinated to four oxygen atoms, and linked to adjacent tetrahedra by sharing three corners to form an infinite two-dimensional hexagonal mesh pattern along the a, b, crystallographic direction. In the octahedral sheet, the octahedra are connected to each other by sharing of edges. The edge-shaped octahedra have hexagonal or pseudo-hexagonal symmetry. Common tetrahedral cations are Si^{4+}, Al^{3+} and Fe^{3+}. Octahedral cations are usually Al^{3+}, Fe^{3+}, Mg^{2+} and Fe^{2+}, but other cations, such as Li^{+}, Mn^{2+}, Co^{2+}, Ni^{2+}, Cu^{2+}, Zn^{2+}, V^{3+}, Cr^{3+}, and Ti^{4+} have also been identified. The 1: 1 layer structure is formed by repetition of tetrahedral and octahedral sheets, while the 2:1 layer structure has one octahedral sheet sandwiched between two tetrahedral sheets (Brigati et al., 2006).

The structure of montmorillonite was first elucidated by Hofmann et al. (1933), and was later modified by Marshall (1935); Maegdefrau and Hofmann (1937); and Hendricks (1942). Montmorillonite is composed of units made up of two silica tetrahedral sheets with a central alumina octahedral sheet. The tetrahedral and octahedral sheets are combined so that the tips of the tetrahedron of each silica sheet and one of the hydroxyl layers of the octahedral sheet form a common layer. The atoms in this layer, which are common to both sheets, are all oxygens instead of hydroxyl. It is thus referred to as three-layered clay minerals with T-O-T layers making up the structural unit (Figure 1). These silica-alumina-silica units are continuous in the 'a' and 'b' crystallographic directions and are stacked one above the other in the 'c' direction. In the stacking of these units, oxygen layers of each unit are adjacent to oxygen of the neighbouring units. This causes a very weak bond and an excellent cleavage between the units. Montmorillonite generally occurs in equi-dimensional, extremely thin flake-shaped units; whereas elongates and needle-shaped units are also common. The probable cause of these elongated structures may be due to the directional strain of the lattice, which occurs due to substitutions within the lattice of ions of a size, which do not fit into octahedral positions (Grim, 1962). In montmorillonite, the symmetry of the layers is C_2 rather than C_2/m (Mering and Glaeser, 1953).

The outstanding feature of this clay structure is that water and other polar molecules can enter between the layers causing the lattice to expand in the 'c' direction. The montmorillonite does not possess a fixed 'c'- axis direction, but can vary from 9.6 Å; the variations depend on the absence or presence of interlayer molecules. The 'c'- axis spacing also varies with the nature of the interlayer cation present between the silicate layers.

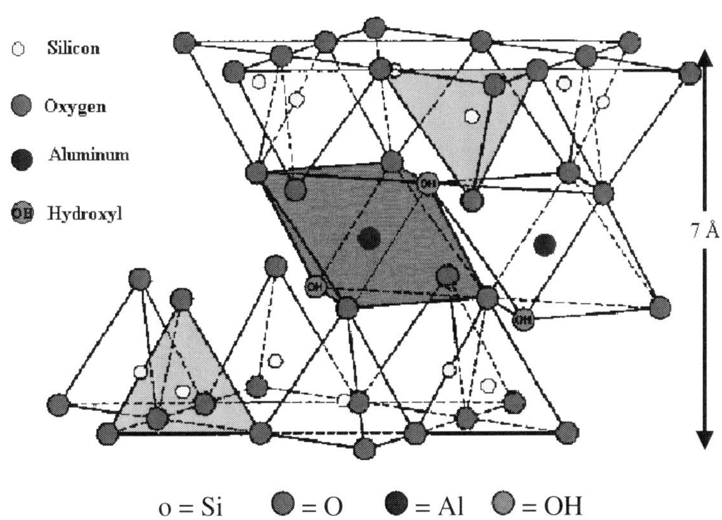

Figure 1. Structure of montmorillonite.

The thickness of the water layers between the silicate units depends on the nature of the exchangeable cations at a given water vapour pressure. Under ordinary conditions, one molecular water-layer and a 'c'- axis spacing of about 12.5 Å occur in montmorillonite having Na as the exchangeable ion. With Ca, there are frequently two-molecular water layers and a 'c'- axis spacing of about 15.5 Å. The expansion capacities are reversible unless the structure completely collapses by the removal of all the interlayer polar molecules. Experiments by Mering (1946), and Bradley and Grim (1948), with montmorillonite in the presence of large quantities of water suggest that with certain adsorbed cations, e.g., Na^+, the unit layers completely separate but that with other cations, for e.g., Ca^{2+} and H^+, the separation is not complete.

The formula for montmorillonite is, $(Si_{7.8}\ Al_{0.2})^{IV}(Al_{3.4}\ Mg_{0.6})^{VI}O_{20}(OH)_4$ and the theoretical composition without the interlayer material is SiO_2, 66.7%, Al_2O_3, 28.3%, H_2O, 5%. The above formula indicates that there is substitution for Si^{4+} by Al^{3+} in the tetrahedral layer and for Al^{3+} by Mg^{2+} in the octahedral layer. Thus, the net layer charge of montmorillonite is: [7.8 (+ 4)] + [0.2 (+ 3)] + [3.4 (+ 3)] + [0.6 (+ 2)] + [20 (−2)] + [4 (−1)] = − 0.8 charge/unit cell. The resulting positive net charge deficiency is balanced by exchangeable cations adsorbed between the unit layers and around their edges. Na^+, K^+, Ca^{2+} and Mg^{2+}, which balance the negative 2:1 layer charge, are commonly hydrated and exchangeable. In the interlayer space, two types of hydration complexes can form: (i) inner-sphere complexes, where the cation is directly bound to the clay surface on one side and to a number of water molecules on the other side, (ii) outer-sphere complexes, where the interlayer cation is completely enfolded by water molecules and interacts with the clay surface through the aqua ligands (Brigati et al., 2006). The electrostatic attractions between the layers and the interlayer cations increase the stacking order in highly charged 2:1 montmorillonite (Lagaly, 2006).

Organo-Montmorillonite

Neutral organic ligands can form complexes with the interlayer cations of the clay minerals (Lagaly et al., 2006). The hydrophilic nature of clay surface changes to hydrophobic by replacement of inorganic interlayer cations by cationic surfactants (Lagaly et al., 1983). This hydrophobic nature of the organo-clay makes it a good filter medium for water purification (Xi et al., 2009). It has been reported that montmorillonite clay can adsorb the organic ions, giving rise to basal spacing greater than those of the original clay minerals saturated with smaller cations such as Ca^{2+} (Lee and Tiwari, 2012).

Modification of montmorillonite with quaternary ammonium cations (QACs) of the form $[(CH_3)_3NR]^+$ or $[(CH_3)_3NR_2]^+$, namely TMA (tetra-methylammonium), TMP (tetra-methylphosphonium), TMPA (trimethyl-phenylammonium), TBA (tetra-butylammonium) cations had been reported (Stevens and Anderson, 1996; Lawrence et al., 1998, Bhattacharyya and Sen Gupta, 2009). Akcay et al. (2006) used N, N′-didodecyl -N, N′- tetramethylethanediammonium (DEDMA) for modification of montmorillonite. In another work, Liu et al. (2012) used three quaternary ammonium cations with different alkyl-chain lengths as modifiers for montmorillonite. The cation exchange capacities of the organo-montmorillonites decreased with QACs chain length and increased loading. Other quaternary ammonium cations, namely, hexadecyl-trimethylammonium bromide, poly(ethylene glycol) (Koyuncu et al., 2011), cetylpyridinium chloride (Sun et al., 2013), cetyl-trimethylammonium bromide (Jing et al., 2013), octadecyl- trimethylammonium and hexadecyl-trimethylammonium cations (Bianchi et al., 2013), etc. were used for this purpose.

During the last few years, various other organic species, namely, 8-hydroxy quinoline (Gök et al., 2008), 3-mercaptopropyltrimethoxysilane (Carvalho et al., 2008), poly(methoxyethyl)acrylamide (Śölener et al., 2008), 2,2'-dipyridyl (Erdem et al., 2009), 2-aminomethyl pyridine (Guerra et al., 2010), 3-(2-aminoethylamino) propyltrimethoxysilane (Balomenou et al., 2008), etc., have been used to modify montmorillonite.

DYES IN THE ENVIRONMENT

Dyes are ionic, aromatic organic compounds with structures including aryl rings having delocalised electron system. Dyes are classified as anionic (acid, direct and reactive dyes), cationic (basic dyes) and non-ionic (disperse dyes). The chromophores in anonic and nonionic dyes mostly consist of azo groups or anthraquinone types. The fused aromatic structures of anthraquinone based dyes make them much more resistant to degradation (Srinivasan andViraraghavan, 2010). Acid dyes are water-soluble anionic dyes with different chromophore groups substituted with acidic functional groups such as nitro-, carboxyl-, and sulfonic acid. The water-insoluble dye becomes soluble by adding a sulfonic group. Direct dyes are highly water-soluble salts of sulfonic acid of azo dyes. Reactive dyes are also highly water-soluble anionic dyes, but have excellent wet fastness with binding to the textile fibers via covalent bonds. Basic dyes are cationic types with chromophores typically having amino groups. Disperse dyes are substantially water-insoluble nonionic dyes for the application to the hydrophobic fibers from aqueous dispersion (Hao et al., 2000).

Dyes are considered an objectionable type of pollutants because they are generally toxic when inhaled or ingested orally. Dyes also cause skin and eye irritation. The main concerns involve the adverse effects of dyes to the environment, including their inhibitory effect on aquatic photosynthesis, ability to deplete dissolved oxygen, and toxicity to flora, fauna and humans. The self-purification ability of streams and conventional biological treatment systems are hindered by the presence of dyes and if the dyes are broken down anaerobically, aromatic amines are generated, which are very toxic, carcinogenic and mutagenic. They are complex aromatic compounds with significant structural diversity. Further, the dyes have a tendency to sequester metals and may cause microtoxicity to fish and other organisms (Crini, 2006; Rangabhashiyam et al., 2013).

The coloured dye wastewater arises as a direct result of the production of the dyes and also as a consequence of their use in the textile and other industries (Allen and Koumanova, 2005). The wastewater from these industries is characterized by high alkalinity, biological oxygen demand (BOD) and chemical oxygen demand (COD) (Kaushik and Malik, 2009). Often, dyes are organic molecules that cause strong colour in the wastewater. The synthetic origin and complex aromatic structures of dyes make them stable and difficult to be biodegraded (Fewson, 1998).

Phenols and Substituted Phenols in the Environment

Phenol is a product of combustion of coal, wood, municipal solid waste and auto exhaust. Pulmonary congestion, interstitial edema, renal tubular hemorrhage and inflammatory changes in the lungs are observed in ingestion of phenol. Extensive renal autolysis is also observed in phenol poisoning. Application of phenol to the skin causes edema, erythema, necrosis, and even death (ATSDR, 2008).

Chlorophenols are a group of chemicals in which chlorines have been added to phenol. Chlorophenols have been recognized as hazardous substances, because most of them are toxic and persistent in the environment. These can enter the environment while they are being made or used as pesticides. Chlorophenols are also the by-products of pulp mill effluent chlorination. It has been suggested that individual chlorophenols should not be present in drinking-water at a level above 0.0001 mg L^{-1} for organoleptic reasons (WHO, 2003). The International Agency for Research on Cancer (IARC) has determined that the chlorophenols as a group are possibly carcinogenic to man (ATSDR, 1999). Putting chlorophenols on the skin or eyes of animals causes severe injuries which is greatest with exposure to the mono- and dichlorophenols. Monochlorophenols can damage the cornea of the eyes.

Another group of phenol analogues, the nitrophenols are commonly used in the manufacture of wood preservation agents, pesticides, dyes and phenolic resins. They are also the breakdown products of several pesticides including parathion and fluoridifen. Their presence has also been detected in auto exhaust.

Adsorption: A Technique for Wastewater Treatment

A number of water treatment technologies are available for removal of contaminants, including dyes from solution. The process of adsorption has become one of the preferred methods for removal of toxic contaminants from water as it has been found to be very effective, economical, versatile and simple (Tran et al., 1999). It is a better technique due to its ease of operation and insensitivity towards toxic substances (Meshko et al., 2001). The adsorption process involves an array of phenomena that can alter the distribution of pollutants among the constituent phases and interfaces of subsurface system (Weber et al., 1991), and convey the separation of a substance from one phase accompanied by its accumulation at the surface of another. The unbalanced forces allied with surface molecules of adsorbents or adsorbates are responsible for the adsorption process. Adsorption phenomena involve different types of attractive forces between adsorbate, adsorbent and solvent molecules. Such forces usually act in concert, but one particular type may be more considerable in any particular situation.

Surface area and the distribution of area with respect to pore size of an adsorbent are two very important factors in determining the extent of adsorption. The nature of the intra-particle surface area markedly influences the types of adsorption process. The adsorbent must have good mechanical properties such as strength and resistance to attain slow destruction. How quickly the adsorbing molecules are transferred to the adsorption sites is another important criterion of an adsorbent (Thomas and Crittenden, 1998). The physical size and form of the adsorbent particles also influence the extent of adsorption. The chemical property of an adsorbent, namely, degree of ionization at the surface of the adsorbent, the types of functional groups present, and the degree to which these properties change in contact with the solution are important considerations in determining the adsorption capacity of a solid.

The extent of adsorption relates to certain properties of the adsorbate relative to the solution phase, namely surface tension and solubility. The extent of adsorption is influenced by the solvophobicity or lyophobicity of a component in the solvent phase. Adsorption from aqueous solution increases as a homologous series is ascended, largely because the expulsion of increasingly large hydrophobic molecules from water permits an increasing number of water-water bonds to reform. If the molecular size of the adsorbate is too large, adsorption will be hindered resulting in decreasing adsorption capacity as very large molecules block or cannot penetrate pores or pathways within the adsorbent. Ionization of the adsorbate in solution affects the extent of adsorption. Many organic compounds ionize in water under appropriate pH altering certain physical and chemical properties, and thus affecting adsorbability. Moreover, a polar solvent can more strongly adsorb on a polar solute from a nonpolar solvent.

The solution pH is one of the important parameters that influence the adsorption process (Mattson and Mark, 1971). The process can also be remarkably influenced by the temperature. The presence of competing adsorbate molecules often influences the extent of adsorption. Few adsorbents demonstrate controllable selectivity for specific adsorbates and thus, all adsorbable compounds present will compete for adsorption sites.

During the last few years, several reviews have appeared on water treatment through adsorption. Bhatnagar et al. (2010) used coconut based adsorbents for removal of dyes and other pollutants, like phenol and substituted phenols, inorganic pollutants, etc. The authors highlighted the development of the preparation of adsorbents utilizing coconut wastes and its major challenges together with the future prospective covering about 150 references. Adsorptive removal of various water pollutants, viz, heavy metals, dyes, phenols, etc., on low cost adsorbents, namely, alumina, silica gel, zeolite, resin, activated carbon, natural materials like wood, peat, coal, lignite, etc.; industrial/agricultural/domestic wastes or by-products such as slag, sludge, fly ash, bagasse fly ash, red mud, etc., were described in another review (Gupta et al., 2009). These adsorbents were of low cost, available in nature and can be generated from waste. In another review, a comprehensive account (with 320 references) was given on the detoxification of water and wastewater by low-cost adsorbents utilizing agro-industrial and municipal wastes (Bhatnagar and Sillanpää, 2010). The use of natural and modified zeolites for removal of heavy metals, inorganic anions and dyes was also reviewed (Wang and Peng, 2010). Modification of natural zeolite was carried out by acid treatment, ion exchange, and surfactant functionalisation process. Adsorption of phenols on low-cost adsorbents like clay, siliceous materials, zeolites, chitin and chitosan, peat and various waste materials / by-products from agriculture and industry was reviewed by Ahmaruzzaman (2008) with a critical analysis of the materials with respect to their characteristics, advantages and limitations and the mechanisms of adsorption. The technical feasibility of the use of activated carbon, synthetic resins, and various low-cost natural adsorbents for the removal of phenol and its derivatives from contaminated water has been reviewed by Lin and Juang (2009).

Due to the vast availability, low cost and eco-friendly nature, the clay minerals got immense importance in wastewater treatment. The adsorption properties of the raw clays, activated clays (acid-treatment or calcinations), organic-modified clays (with small molecules or polymers) for removal of organic dyes from aqueous solutions were discussed by Liu and

Zhang (2007). The authors reported that though the nano size of clay minerals favour the adsorption process, the difficulties in the dispersing or separating procedures of clays do not allow this advantage. Yu et al. (2013) summarized adsorption of proteins and nucleic acids (DNA, RNA) on natural clay minerals. The interaction mechanisms were suggested to be cation exchange, electrostatic interactions, hydrophobic/hydrophilic interactions, ligand exchange, Cation Bridge, Water Bridge, hydrogen bond and van der Waals forces. The use of kaolinite and montmorillonite clays for removal of heavy metals, like, Cd, Cr, Co, Cu, Fe, Pb, Mn, Ni, Zn, etc., were also reviewed (Bhattacharyya and Sen Gupta, 2008; Sen Gupta and Bhattacharyya, 2012). The adsorption capacities varied from metal to metal and also depended on the type of clay used. Moreover, the experimental conditions, viz., clay amount, initial metal ion concentration, stirring speed, solution pH, temperature, etc., influenced the adsorbate–adsorbent interactions. Very recently, the same group of authors reported an extensive survey of adsorption of metal ions on clays and some inorganic solids covering the period, 2000–2013 (Sen Gupta and Bhattacharyya, 2014). The study described the use of isotherms to obtain the adsorption capacities of the solids.

In the present review, preparation and characterization of the organo-bentonites and their use as adsorbents for dyes and other organic pollutants, namely, phenols and substituted phenols in water are dealt with based on the published work during the last ten years.

THEORETICAL BASIS OF ADSORPTION

Adsorption Isotherm

Experimental isotherms are useful for describing adsorption capacity to find out the viability for the process for a given application, for selection of the most appropriate adsorbent along with its required doses. Adsorption equilibrium is established when the concentration of the adsorbate in the bulk solution is in dynamic balance with that of the interface. Although several isotherm models are in use, Freundlich and Langmuir models have found the widest use and are briefly introduced below.

Freundlich model (Freundlich, 1906) of non-linear adsorption equilibrium is used for obtaining the adsorption capacities of different solids. The isotherm has the expression:

$$q_e = K_f C_e^n \tag{1}$$

where C_e and q_e are the equilibrium concentrations of the metal ions in the liquid phase and in the solid phase respectively, K_f and n being Freundlich coefficients, representing adsorption capacity and adsorption intensity respectively. For favourable adsorption, n is generally < 1. The logarithmic form of the equation (1) is more useful for fitting data from batch equilibrium studies, and is given by

$$\log q_e = \log K_f + n \log C_e \tag{2}$$

The values of K_f and n are obtained from intercept and slope of the plots of log q_e vs log C_e, respectively.

The Langmuir model (Langmuir, 1918) is the most widely used non-linear isotherm model and is very frequently used to evaluate the efficiencies of various solid adsorbents. Langmuir capacities provide a useful and convenient method to compare the adsorption capacities of similar or different types of solids. The equation is based on evaluating the fractional surface coverage, q

$$\theta = q_e/q_m = bC_e/(1+bC_e) \tag{3}$$

where q_m is the quantity of adsorbate required to form a single monolayer on unit mass of the solid, popularly known as the Langmuir capacity, q_e is the amount adsorbed at equilibrium on unit mass of the solid, C_e is the concentration of gas or solute remaining unadsorbed at equilibrium, and b is Langmuir coefficient related to the adsorption equilibrium constant. Equation (3) is usually written in the simplified form,

$$C_e/q_e = (1/bq_m) + (1/q_m)\, C_e \tag{4}$$

The linear Langmuir plots are obtained by plotting C_e/q_e vs. C_e. The slope and the intercept of this plot give the values of q_m and b.

Adsorption Kinetics

Kinetics is the study of the rates of chemical processes to understand the factors that influence the rates. The adsorption kinetics normally includes two phases: a rapid removal stage followed by a much slower stage before the equilibrium is established (Ho and McKay, 1999a).

Lagergren Pseudo First Order Model

The Lagergren equation (Lagergren, 1898; Ho, 2004) is probably the earliest known example describing the rate of adsorption in the liquid-phase systems. The equation is given by,

$$dq_t / dt = k_1 (q_e - q_t) \tag{5}$$

where $k_1(\text{min}^{-1})$ is the pseudo first order adsorption rate constant. The integrated form of the equation (5) for the boundary conditions of $t = 0$, $q_t = 0$ and $t = t$, $q_t = q_t$, is

$$\ln (qe - qt) = \ln qe - k1t \tag{6}$$

where q_e and q_t are the values of the amount adsorbed per unit mass at equilibrium and at any time t. The values of k_1 can be obtained from the slope of the linear plot of ln $(q_e\text{-}q_t)$ vs. t.

An important test for the validity of the pseudo-first order kinetics is to compare q_e obtained from the intercept (ln q_e) of the plots of ln $(q_e\text{-}q_t)$ vs. t with the value obtained experimentally when the adsorption reaches equilibrium. In most cases, the Lagergren equation does not fit well in the whole range of interaction time (Ho and McKay, 1999b) and then, another suitable model will have to be looked for.

Pseudo-Second Order Equation

When the applicability of the first order kinetics becomes untenable (for example, if q_e obtained on the basis of equation (6) is not the same as the one obtained experimentally), the adsorption may follow second order kinetics on the basis of the equation (Ho and McKay, 1999b)

$$dq_t / dt = k_2 (q_e - q_t)^2 \tag{7}$$

where k_2 is the second order rate constant. The integrated form of the equation (7) under the boundary conditions of $q_t = 0$ at $t = 0$ and $q_t = q_t$ at $t = t$ is given by,

$$q_t = t/[(1/ k_2 q_e^2) + t /q_e] \tag{8}$$

The linear form of this equation

$$t/q_t = 1/(k_2 q_e^2) + (1/q_e) .t \tag{9}$$

makes it possible to obtain q_e and k_2 (g $mg^{-1}min^{-1}$) from the plots of t/q_t vs. t.

Experimental Insight into Chemistry of Organo-Montmorillonite

Preparation and Characterization of Organo-Montmorillonite

A series of surfactant-modified montmorillonites (MMT) were prepared using octyltrimethylammonium bromide (OTAB), dodecyltrimethylammonium bromide (DTAB), cetyltrimethylammonium bromide (CTAB) and stearyltrimethylammonium bromide (STAB) by Wang and Wang (2008a). Each of the surfactant (equivalent to CEC of 1.0 of the purified MMT) was dissolved in 100 mL distilled water and then 4.0 g purified clay was slowly added. After continuous stirring (at room temperature) for 12 h, the reaction mixture was filtered and washed several times with distilled water to free it of bromide ions. The resultant modified montmorillonite was dried in an air oven (at 105oC) for 4 h and ground to 200 mesh size. Following the same procedure, a series of CTAB-MMT clay was synthesized except that CTAB with 0.25, 0.5, 0.75, 1.0, 1.5, 2.0, 2.5 and 3.0 CEC of montmorillonite were added to the reaction system. The authors reported that, the IR absorption band at 3442 cm^{-1} (corresponding to –OH stretching vibration of H_2O of MMT) weakened and shifted to the lower wave numbers of 3425, 3426, 3428 and 3430 cm^{-1} for octyltrimethylammonium bromide-, dodecyltrimethylammonium bromide-, cetyltrimethylammonium bromide- and stearyltrimethylammonium bromide-MMT, respectively. The modified MMT possessed symmetric and asymmetric stretching vibrations of $-CH_2$ (2851 cm^{-1}) and $-CH_3$ (2921 cm^{-1}). The increase in the number of carbon atoms of surfactants enhanced the intensity of the bands. Thus, the hydrophilic nature of MMT was altered to hydrophobic by the surfactant modification of MMT. The formation of the intercalated nanostructure with CTAB of 0.5 CEC of MMT was indicated by the shift in the XRD band of MMT to lower angle (5.94°), responding to a basal spacing of 14.89Å. The intensity of the peak disappeared with further increase in the amount of CTAB indicating the formation of the exfoliated nanostructure in CTAB-MMT. SEM measurements showed formation of larger particles with coarse porous surface in CTAB-MMT compared to the original MMT particles.

Hexadecyl trimethyl ammonium, octadecyl amine and (octadecyl/hexadecyl) dihydroxyethyl methyl ammonium salts were used to prepare a series of organo-montmorillonites (Zha et al., 2013) byadding 50 g Na-montmorillonite in suspension the surfactant solution followed by stirring for 150 min at 80^0C. The material was filtered, washed, dried and ground to obtain the adsorbent. XRD measurements showed hexadecyl trimethyl ammonium montmorillonite to have an increased interlayer spacing of 2.314 nm, indicating entry of the surfactant ions into the montmorillonite interlayer space. Moreover, the basal spacing of both octadecyl amine- and octadecyl/hexadecyl) dihydroxyethyl methyl ammonium-montmorillonite increased to 2.433 and 2.467 nm, respectively. The difference was due to the size of the surfactant molecules.

Jiang et al. (2002) intercalated a series of modified montmorillonites with hexdecyl-trimethyl-ammonium (HDTMA) bromide, polymeric Fe/hexdecyl-trimethyl-ammonium bromide, polymeric Al/hexdecyl-trimethyl-ammonium bromide and polymeric Fe/Al/hexdecyl-trimethyl-ammonium bromide. The pillaring solution containing either Al or Fe polymeric species was aged for a predetermined period and 15 g clay was added into it under vigorous stirring and kept overnight. The product was washed and mixed with HDTMA bromide in order to form a surfactant-pillared complex on 1:1 molar basis of HDTMA to Al or Fe. The HDTMA-montmorillonites had increased basal spacing and larger N_2 BET surface areas.

Three novel Gemini surfactants [($C_{30}H_{66}N_2Br_2$), ($C_{31}H_{68}N_2C_{12}$) and ($C_{43}H_{92}N_2C_{12}$)] along with cetyltrimethyl ammonium bromide (CTAB) were used to develop organo-montmorillonite (MMT) by microwave irradiation as well as traditional heating method (Liu et al., 2011). The suspension of MMT in 50% (v/v) mixture of isopropanol and water was stirred (24 h) at room temperature, and transferred to a microwave synthesis system (reaction power: 800 W, reaction temperature: 80°C). 10 mL surfactant isopropanol solution was slowly added to the MMT suspension. After vigorously stirring for 1 h, precipitation of the organo-montmorillonite was obtained. In case of traditional heating method, MMT was kept dispersed in 50% (v/v) isopropanol for 24 h after initial vigorous stirring for 30 min. The appropriate dosage of Gemini surfactant [($C_{43}H_{92}N_2C_{12}$)] was added slowly into the suspension at 80°C under stirring. After stirring for 24 h, the products were washed to remove chloride ions, lyophilized and ground. The organo-MMT had expanded interlayer space which was observed to be positively related to the chain length of the surfactants. Microwave heating created large layer spacing in all three organoclays compared to traditional heating. The IR bands at 3623 and 1638 cm^{-1} of organo-MMTs were perturbed and diminished in intensity, indicating electrostatic interactions between OH groups of organo-MMT and quaternary ammonium groups of Gemini surfactant. The structural change from hydrophilic to hydrophobic character of modified clay was revealed by the decrease in the intensity of the

IR bands. TEM measurements also showed enlarged interlayer distance of the three organo-MMTs. TG and DTG measurements were used to show that the organoclays had lower mass loss than MMT below 200^0C showing less amount of free water in the organoclays. The surfactants were shown to reduce the surface energy of the clay and convert the hydrophilic silicate surface to hydrophobic.

A series of dual-cationic surfactant-modified organobentonites was synthesized by reacting bentonite with aqueous solutions containing quaternary ammonium cations (Ma et al., 2011). 250 ml Hexadecyltrimethylammonium bromide (HDTMAB) solution (100% of CEC) was added dropwise to 10 g bentonite suspension under stirring. 250 ml of other cationic surfactant solution [HDTMAB, myristyltrimethylammonium bromide (MTMAB) or dodecyltrimethylammonium bromide (DTMAB) or octyltrimethyl ammonium bromide (OTMAB)] was added drop wise. The treated bentonites were separated from water, washed, dried (80–90^0C) and mechanically ground to less than 100 mesh. FTIR bands of 2919 and 2851 cm^{-1} corresponded to symmetric and asymmetric –CH_2 stretching vibrations while bands at 1472 and 1418 cm^{-1} assigned to bending vibrations were observed only in HDTMA bentonite.

Hexadecyltrimethyl ammonium bromide (HDTMAB) and octadecyltrimethyl ammonium bromide (ODTMAB) modified montmorillonite were synthesized (Yildiz et al., 2005) by agitating the clay mineral with ODTMA- and HDTMA- solutions for 12 h at 25 oC, followed by separation, washing and drying. It was observed that the basal spacing of organobentonite increased from 12.40 to 17.60 Å, corresponding to the formation of bilayers and paraffin complexes in which ODTMA and HDTMA ions were in direct contact with each other. The BET surface area of the modified clays decreased from 55.37 m^2g^{-1} to 28.92 and 35.68 m^2g^{-1} for ODTMA-bentonite and HDTMA-bentonite, respectively. The modification of bentonite by surfactants converted the original clay into more aggregated forms. The relatively low surface area of the organobentonites indicated only the external surface.

In another work, cetyldimethylbenzylammonium chloride-bentonite (CDBA-bent) and cetylpyridiniumchloride-bentonite (CPC-bent) were synthesized from Na-bentonite (Baskaralingam et al., 2006). The Na-bentonite was prepared by stirring the bentonite with 1N NaCl for 24 h, washed and dried. 0.1M aqueous solution of CDBA and CPC was added dropwise to Na-bentonite suspension and stirred for 24 h. The complex was centrifuged, washed, dried and ground to 200 mesh size. The basal spacing increased from 14.38 Å (bentonite) to 24.65 Å and 22.63 Å for CDBA-bentonite and CP-bentonite respectively. The increase was explained as due to the intercalation of cationic surfactants into the interlayers of bentonite with a monolayer arrangement.

Arellano-Cárdenas et al. (2013) intercalated hexadecyltrimethylammonium bromide into a Mexican montmorillonite. The organo-clay was obtained by adding amounts of surfactant equivalent to 100% of CEC of the host clay. XRD data showed a basal spacing of 2.012 nm. The BET specific surface area, average pore diameter and point of zero charge were reported to be 9.3 m^2 g^{-1}, 11.35 nm and 8.0, respectively.

To prepare hexadecyltrimethyl ammonium-bentonite, Salgin et al. (2007) dissolved an amount of HDTMA-bentonite equal to the CEC of sodium (65 meq/100 g clay) bentonite in 1 L of water and added the same to a bentonite suspension. The mixture was agitated for 12 h and HDTMA-bentonite suspensions were separated, washed and dried.

Hexadecyltrimethyl ammonium-bentonite was also prepared by adding 1 L of the surfactant solution, containing the surfactant equivalent to 100% of the CEC of the clay to a suspension of 50 g of Jordanian bentonite under stirring (Rawajfih and Nsour, 2006). The modified clay was subsequently washed to maintain the pH at 5.5-60 and dried. In a similar way, hexadecyltrimethyl ammonium (HDTMA)-bentonite was prepared by Anirudhan and Ramachandran (2006; 2014) by introducing HDTMA cations into the exchange sites by exchanging Na^+ ions of Na-bentonite clay. 20 g of Na-bentonite was treated with HDTMA solution (the surfactant amount was equal to 100% of the CEC of bentonite). The mixture was stirred for 6 h at 60 oC, filtered, washed several times with distilled water, and dried at 70 oC for 24 h. The product was sieved and particles having the average diameter of 0.096 mm were used as adsorbent. The basic structure of the aluminosilicate layer was shown to have been maintained even after the modification while interlayer space was increased from 16.12 to 34.11 Å. The increase in the basal spacing in the organo-bentonite was suggested to be due to replacement of inorganic interlayer cations and their hydration water with $HDTMA^+$ cations. The increases in the basal spacing indicated to a bilayer arrangement for the surfactant cations. The BET surface area was found to be 36.5 and 27.9 m^2/g for Na-bentonite and organoclay, respectively. The higher surface area of Na-bentonite was shown to be due to the enhanced adsorption of N_2 in the wider micropores and mesopores.

In an another work, the modification of bentonite by hexadecyltrimethylammonium bromide (HDTMAB) was carried out by mixing clay water dispersion and HDTMAB in a microwave oven for 5 min at 360 W (Elemen et al., 2012). The basal spacing of organobentonite increased to 18.85 Å from 11.95 Å in bentonite indicating the existence of quaternary ammonium cations in the interlayer space. The strong FTIR bands at 2850 and 2922 cm^{-1} of organo-bentonite (assigned to symmetric and asymmetric stretching vibrations of the methyl and methylene groups) and their bending vibrations between 1468 and 1475 cm^{-1} supported the intercalation of surfactant molecules between the silica layers.

HDTMA-bentonite was prepared by Mirmohamadsadeghi et al. (2012) from acid treated bentonite. Acid-treatment increased the basal spacing of the natural bentonite from 12.6 to 14.5 Å and the basal spacing of surfactant modified bentonite and surfactant modified acid-treated bentonite increased to 19.2 and 19.0 Å respectively, indicating the intercalation of the surfactant into the clay layers. Koswojo et al. (2010) synthesized organo-bentonite by mixing 10 g bentonite in 100 mL cetyl trimethylammonium bromide (CTAB) (150 % of CEC) in a microwave oven (600 W, 10 min). Organo-bentonite possessed larger volumetric adsorption capacity (0.70 cm^3/g) and lower surface area (96 m^2/g) than the parent bentonite (0.48 cm^3/g, 118 m^2/g) indicating the formation of larger pore on organo-bentonite due to CTAB insertion at bentonite interlayer. The modification of bentonite by CTAB increased the basal spacing from 12.3Å to 25.1Å and making the clay surface hydrophobic. FTIR bands at 2819.97 cm^{-1} and 2920.40 cm^{-1} were associated with symmetric and asymmetric stretching vibrations of CH_2 and CH_3 groups of the aliphatic chain in the surfactant. The intercalation of the surfactant between the silica layers was confirmed by the presence of the band, 1475.19 cm^{-1}, due to the bending vibration of the methylene groups.

Cetyltrimethylammoniumbromide (CTAB) modified bentonite was also reported by Bhatt et al. (2012) by suspending a known mass of clay in an aqueous solution of CTAB (CTAB was 100% of the CEC of clay). The suspension was stirred at room temperature for 24 h and the product was separated, washed, dried and ground. The presence of CTAB in the interlayers increased the basal spacing from 12.7 to 19.5 Å. Zohra et al. (2008) also reported preparation of cetyltrimethylammonium modified bentonite by dissolving the surfactant in 1 L of water and adding 10 g of the clay. After agitation for 24 h, the modified clay was recovered by centrifugation, washed and dried. The increase in the basal spacing was from 12 to 19Å in the modified clay indicating intercalation of the cationic surfactant into the interlayers of bentonite with a bilayer arrangement, giving a thickness of about 7Å. The IR absorption band at 1468 cm^{-1} represented the bending vibrations of NH_4^+ and their bonding supporting the intercalation of surfactant molecules between the silica layers. In yet another work, cetyltrimethylammonium-bentonite was prepared by adding 5.0 g bentonite dispersion (200 mL water) to the surfactant solution and stirring for 12 h (at 70^0C) (Zhu and Zhu (2008), Both BET surface areas and total pore volumes decreased by introduction of the surfactant cations with the values reversely related to the surfactant load. With increasing surfactant load, the surface fractal dimension gradually decreased indicating a more regular organo-bentonite surface, attributed to the filling of the pores by the surfactant cations.

Senturk et al. (2009) synthesized CTAB-bentonite by first adding 1M Na_2CO_3 solution to Ca-bentonite and stirring for 3 h at 800 rpm, the solid Na-bentonite was dried and mixed with 200 mL 4% CTAB solution and 20 g Na–bentonite by stirring for 24 h. IR bands at 2920–2850 cm^{-1} represented $–CH_3$ and $–CH_2$ stretching frequencies and the band near 1460 cm^{-1} was associated with $–CH_2$ deformation frequency in line with the modification of bentonite by CTAB. SEM revealed considerable numbers of heterogeneous pores in the organobentonite.

Xin et al. (2011) reported the synthesis of octadecyl trimethyl ammonium chloride (OTMAC) and octadecyl trimethyl ammonium chloride/hydroxy-aluminum (Al(OH)-OTMAC) modified bentonite. To prepare OTMAC-bentonite, 20 g Na-bentonite was shaken with OTMAC-solution for 24 h. The modified bentonite was washed, dried and ground to pass through a 200 mesh sieve. To prepare Al(OH)-OTMAC-bentonite, the pillaring solution of hydroxy-aluminum cations and the OTMAC-solution were added drop-wise into bentonite suspension with stirring. The products were washed, dried and ground to pass through a 200 mesh sieve. XRD showed that the basal spacing increased to 20.22 and 20.73 Å for OTMAC-bentonite and Al(OH)-OTMAC-bentonite respectively, larger than that of the parent bentonite. The BET surface areas increased from 10.15 m^2g^{-1} to 26.70 and 71.54 m^2g^{-1} for OTMAC-bentonite and Al(OH)-OTMAC-bentonite respectively.

In another work, organo-bentonites were synthesised by using cetyltrimethyl-ammonium bromide (CTMA), tetramethylammonium bromide (TMA) and the dye, crystal violet dye (CV) (Huang and Zhu, 2011). CTMA- and CV- modified bentonites possessed much larger basal spacings than TMA-bentonite, due to the larger molecular sizes of CTMA and CV. In addition, CV- and TMA-synthesized organo-bentonites had quite similar basal spacings despite their different charges.

Özcan et al. (2004) reported the preparation of dodecyltrimethylammonium bentonite from Na-bentonite. The dodecyltrimethylammonium bromide (equal to twice the CEC of bentonite) was added to 30 g Na-bentonite dispersed in 0.5 L of distilled water. The final product was separated, washed and collected. The symmetric and asymmetric stretching vibrations of the methylene group created a pair of strong bands at 2856 and 2929 cm^{-1}. Moreover, the intercalation of the surfactant molecules between the silica layers was supported by appearance of the bending vibrations between 1396 and 1486 cm^{-1}.

Xu and Zhu (2009) used two surfactants with different chain structure (octadecyltrimethylammonium chloride, OTMAC and dioctadecyldimethylammonium chloride, DODMAC) to synthesis a series of organo-bentonites. To 0.2 g bentonite dispersion (20 ml water), OTMAC solution (0.2–1.4 times the CEC of bentonite) and DODMAC solution (0.1–0.7 times the CEC of bentonite) were added and shaken for 24 h at 25 oC. The modified bentonite was, dried and pulverized to pass through an 80 mesh sieve. The basal d spacing of OTMA-modified clay increased with surfactant loading [1.50 nm (0.2 CEC), 1.58 nm (0.4–0.8 CEC), 2.40 nm (0.6–1.2 CEC) and 4.16 nm (1.0–1.4 CEC)]. At low surfactant loadings (0.2–0.8 CEC), the $OTMA^+$ chains lay parallel to the bentonite planes forming flat monolayers or lateral-monolayers. With increasing surfactant loading (0.6–1.2 CEC), pseudo-tri-molecular layer structure was shown to become the predominant conformation. The interlayer of 1.0–1.4 CEC OTMA-Bentonite had d-spacing of 3.20 nm, suggesting that the $OTMA^+$ arrangement might be a paraffin type bilayer. For DODMA-bentonite, d spacings of 1.51 nm (0.1–0.5 CEC) and 3.71 nm (0.5–0.7 CEC) were observed. At low surfactant loadings, a flat monolayer arrangement was the possible conformation for $DODMA^+$ chains in the modified clay. The two alkyl chains linked to the same N-atom of $DODMA^+$ lay parallel to the silicate plane. At high surfactant loadings, a paraffin-type bilayer to the silicate planes was the predominant arrangement of intercalated $DODMA^+$. IR data revealed that the conformations of sorbed $OTMA^+$ and $DODMA^+$ transformed from disorder (more gauche conformation) to order (more trans conformation) with increasing surfactant loading and the packing density of the alkyl chains became higher. This was further supported by XRD data.

Na^+-exchanged bentonite was also used as the starting material for two series of organobentonites (Jović-Jovičić et al., 2013). The first series consisted of organobentonites with different hexadecyltrimethylammonium (HDTMA) loadings. The second series was obtained using hexadecyltrimethylammonium (HDTMA), dodecyltrimethylammonium (DDTMA) and tetramethylammonium (TMA) ions having different chain lengths in varying amounts. Initially, Na^+-exchanged bentonite was prepared by stirring 10 g of raw bentonite dispersed in 1.0 L of 1 M NaCl for 24 h thrice. The appropriate surfactant solution (150 ml) was added to 50 g Na-bentonite in 100 ml distilled water and stirring for 24 h at room temperature. The synthesized product, after separation, washing, drying and grinding, gave XRD data revealing that increase in alkyl chain length from TMA- to HDTMA- increased the basal spacing. However, the increase in HDTMA after a certain optimum amount could not increase the basal spacing, which indicated that after replacing all the exchangeable interlayer cations, additional HDTMA ions were attached to the outer clay surface. The increase in BET surface area and pore volume for 2.0 TMA-clay was mainly caused by the presence of newly developed pores, as a consequence of the incorporation of short-alkyl chain TMA in the form of pillars. On the other hand, the presence of cations, namely, HDTMA (with 16 C atoms) and DDTMA (with 12 C atoms) in the interlamellar region of bentonite resulted in pore blockage that inhibited the passage of nitrogen molecules, resulting in lowering of BET specific surface area.

In another study, Wibulswas (2004) modified montmorillonite by incorporating tetramethylammonium chloride (TMA), tetradecyltrimethylammonium bromide (TDMA), hexadecyltrimethylammonium bromide (HDTMA), and benzyldimethylhexadecylammonium chloride (BDHDMA). The d_{001} peak of most of the modified samples shifted to lower diffraction angle compared to the parent clay, showing that the binding of organic cations created a hydrophobic environment in the interlayer space of the clays. The amount and the alkyl chain length of the quaternary ammonium compounds (QAC) influenced the interlayer spacing in the order TMA-montmorillonite < TDMA-montmorillonite < HDTMA-montmorillonite < BDHDMA- montmorillonite. The modified clays had larger particle size than montmorillonite and the particle size increased as a function of the amount and the molecular size of the QACs.

Alkaram et al. (2009) prepared hexadecyltrimethylammonium (HDTMA)–bentonite and phenyltrimethylammonium (PTMA)–bentonite. 50 mmol L^{-1} aqueous HDTMA or PTMA solution was added to a 7% aqueous clay suspension and the mixture was stirred for 3 h. The modified clay was separated, washed and dried at 110^0C. The IR spectra of HDTMA-bentonite showed a pair of strong bands at 3434 and 3625 cm^{-1} due to symmetric and asymmetric stretching vibrations of the methylene groups, and a band between 1384 and 1470 cm^{-1} due to bending vibrations of the same groups, indicating entry of HDTMA molecules between the silica layers. The appearance of an absorption band at 693 cm^{-1} was attributed to mono-substituted benzene group of PTMA.

Phenyltrimethylammonium (PTMA)- and benzyltrimethylammonium (BTMA)-bentonite were prepared by using Na-bentonite (Majdan et al., 2009) when 1.0 g Na-bentonite was contacted with 0.1 L of 0.01 M aqueous/ethyl alcohol solutions of either PTMA or BTMA at 60 oC for 6 h. The intercalation of PTMA to bentonite was shown to decrease the intensity of the IR bands at 1400–1600 cm^{-1} and the band at 1595 cm^{-1} disappeared. The bands responsible for the phenyl ring vibrations in the region 1400–1600 cm^{-1} shifted to lower frequencies when BTMA and BTMA-bentonite were compared. The authors concluded that the interactions of BTMA- with siloxane oxygens were stronger than that of PTMA.

Akçay (2004) used tetraethylammonium (TEA) iodide for bentonite modification by mixing dried bentonite with the surfactant for 42 h at 25 oC. The treated sample was separated by centrifugation and washed with ethanol and ethanol-water (1:1), dried and mechanically ground to 140 mesh. The basal spacing increased from 9.5 to 14.2Å for the modified clay as observed from XRD spectra. This suggested that TEA ions intercalate into the interlayers of bentonite with a monolayer arrangement. The surface area of the tetraethylammonium-bentonite was found to be 71 m^2/g from BET nitrogen adsorption/ desorption isotherms.

Polydiallydimethylammonium (PDADMA) modified bentonite was synthesized by mixing 100 mL 4% PDADMA solution and a slurry of 20 g bentonite (at 65 oC) and stirring for 3 h. The modified clay was washed, dried and activated for 1 h at 105 oC (Shen et al., 2009). The basal spacing of PDADMA–bentonite increased from 1.28 to 1.51 nm as PDADMA molecules partially intercalated into the interlayers of bentonite and expanded the interlamellar spacing.

In another work, bentonite was modified by introducing epichlorohydrin dimethylamine (EPIDMA) in which 100 ml 5 % poly(epichlorohydrin dimethylamine) solution was stirred with 20 g bentonite dispersion at 70^0C for 2 h (Li et al., 2011). Poly(epicholorohydrin dimethylamine) modified bentonite was also prepared by Kang et al. (2009) by adding slowly 100 ml of 4 % EPIDMA solution to 20 g Na^+-bentonite dispersion and stirring for 2 h at 55 oC. The EPIDMA/bentonite complex was washed, dried, ground and sieved through 300 mesh sieve. XRD patterns of bentonite and EPIDMA-bentonite were found to be similar in the range of 10–40° indicating that EPIDMA molecules only partially intercalate into the interlayer spaces. Due to the influence of the interface constraint of polymer intercalation, the polymer cations could not exchange the interlayer cations completely. The averaged particle size of EPIDMA- bentonite (~ 25 μm) was much larger than that of the raw bentonite (~3 μm). The EPIDMA molecules expanded the interlayer space resulting in an increase in the specific surface area of EPIDMA/bentonite (27.5 $m^2 g^{-1}$) compared to that of the raw bentonite (12.9 $m^2 g^{-1}$).

Polyacrylamide-bentonite composite with amine functionality (Am-PAA-B) was prepared by direct polymerization in the presence of N,N'-methylenebisacrylamide (MBA) as a crosslinking agent and potassium peroxydisulphate as an initiator followed by reaction with ethylenediamine and modification of Am-PAA-B by immobilizing humic acid (Anirudhan et al., 2009). 10 g bentonite was stirred with 1 L (1.0 M) NaCl solution for 12 h, centrifuged, washed and dried. To a homogeneous solution of 10 g Na-bentonite (in 250 ml water), 20 g acrylamide (in 50 ml water) was added and stirred for 2 h, mixed with 1.6 g MBA and 500 mg $K_2S_2O_8$, followed by addition of 0.5 ml NNN'N'N"N"-hexamethylenediammine (HMD) for polymerization and heated in a water bath (60 oC). The polymer-bentonite composite (PAA-B) was washed and dried at 60 oC. To introduce the amine functionality in PAA-B, 10 g PAA-B was heated with 50 ml ethylenediammine at 80°C (pH 11.0 for 4 h). The product, Am-PAA-B was washed and dried. In the next step, 500 ml (1250 μmol L^{-1}) humic acid was added to 10 g Am-PAA-B suspension (in 100 ml water) and stirred for 24 h at room temperature. The product (humic acid immobilized Am-PAA-B) was filtered, washed, dried at 50 oC and sieved with 80–230 mesh sieves.

The immobilization of 2-aminomethylpyridine on smectite was reported by Guerra et al. (2011). In the preparation, 20.0 cm^3 (108.0 mmol) 3-chloropropyltrimethoxysilane was added to 20.0 g smectite suspension (in 100.0 cm^3 dry toluene) and was stirred under reflux of the solvent in an inert atmosphere for 72 h. The suspension was filtered, washed and dried. In the next step, 5.0 g of this modified clay was reacted with 3.50 cm^3 (0.034 mmol) of 2-aminomethylpyridine in 50.0 cm^3 toluene, containing 5.0 cm^3 triethylamine. The suspension was stirred under reflux for 72 h at 363 K.

Diphosphonium-intercalated montmorillonites were prepared via ion exchange using para, meta and ortho-bis(triphenyl phosphonium methylene)-benzene-dichloride (TPhPMB) (Makhoukhi et al., 2010). 10 g montmorillonite was slowly added to 100 mL 0.5 M aqueous solutions of o-, m- or p-TPhPMB and was stirred at room temperature for 24 h.

1,6-diamino hexane (DAH) modified bentonite was prepared by Gök et al. (2010) by taking 30 g natural bentonite dispersed in 0.8 dm^3 water and mixing with the appropriate amount of 1,6-diamino hexane (1.5 times of the CEC of bentonite), followed by stirring for 24 h. The IR stretching vibrations of $-CH_2$ groups at 2951 and 2850 cm^{-1} were observed only for DAH-bentonite indicating modification. The specific surface area of DAH-bentonite was 109.95 $m^2 g^{-1}$ (bentonite: 67.49 $m^2 g^{-1}$). The increase in surface area was induced by the incorporation of 1,6-diaminohexane molecules between the clay layers caused by the presence of the micropores. The modification was confirmed by a DTG peak at 440.81 oC, not observed in natural bentonite. Another DTG peak around 100 oC showed decreased intensity for DAH-bentonite compared to that of natural bentonite due to the increase in hydrophobicity as a result of modification.

Bentonite was also modified with chitosan and hexadecyl trimethyl ammonium bromide (Guo et al., 2012). The chitosan solution was prepared with 1.0% acetic acid with constant stirring and bubbling of nitrogen gas at room temperatures. 0.02 mol L^{-1} hexadecyl trimethyl ammonium bromide was added to 5.0 g bentonite taken in 200 mL of water and stirring vigorously for 2 h. To it, 1.0% chitosan solution was added and stirred for 2 h. A series of biopolymer chitosan/montmorillonite nanocomposites were prepared (Wang and Wang, 2007) by controlling the molar ratio of chitosan and montmorillonite. 4.0 g montmorillonite dispersion (in 100 ml water) was mixed with chitosan solution (0.0660, 0.132, 0.660, 3.30, 6.60 g chitosan in 1% (v/v) aqueous acetic acid) slowly by stirring at 60 oC for 6 h (chitosan to montmorillonite molar ratios of 1:10, 1:5, 1:1, 5:1 and 10:1). The synthesized composites were washed, dried and sieved to 200 mesh size. The typical diffraction peak of montmorillonite (6.94°), corresponding to a basal spacing of 12.74Å, disappeared after intercalation with chitosan. The formation of the flocculated–intercalated nanostructure with the molar ratios of 1:5 was revealed by the movement of the typical diffraction peak of montmorillonite to a lower angle (5.61°). The intensity of the peak decreased and even disappeared with increasing molar ratios of chitosan to clay, indicating the formation of an intercalated–exfoliated structure in chitosan modified clay nanocomposites. The vibration bands in chitosan (O–H and N–H stretching) overlapped with the bands of montmorillonite (–OH stretching of H_2O) as shown by the enhancement of the IR absorption band intensity of the nanocomposites at 3424 cm^{-1}. The bands attributed to the intercalated chitosan (C–H stretching of methyl groups at 2923 cm^{-1} and methylene groups at 2882 cm^{-1}) were observed in the spectra of the nanocomposites, and the intensity of both the bands increased with increasing molar ratio of chitosan to montmorillonite.

The same group of workers (Wang and Wang, 2008b) reported the preparation of N,O-carboxymethyl-chitosan/montmorillonite nanocomposites. 5.0 g chitosan in 20 wt% NaOH solution (50.0 mL) was aged for 12 h and was mixed with 50 mL of anhydrous alcohol by stirring for 30 min at room temperature, followed by addition of 3.58 g monochloroacetic acid and stirring for additional 30 min. The mixture was heated at 200 oC for 1 h. The precipitate (N,O-CMC) was filtered, washed and dried. N,O-CMC solution (containing 0.0185, 0.037, 0.185, 0.925 and 1.85 g N,O-CMC) was added slowly to 1.0 g montmorillonite (in 100 mL water) by stirring at 60 oC for 6 h. The synthesized composites were washed, dried, ground and sieved to 200 mesh size. After intercalation with N,O-CMC, the montmorillonite XRD peak at 6.94^0 (basal spacing of 12.74Å) shifted to a lower angle and even disappeared. The appearance of a diffraction peak at 5.78^0 (d = 1.53 nm) for molar ratio of N,O-CMC: montmorillonite of 1:5 indicated the formation of intercalated nanocomposite. The disappearance of this peak for the molar ratio of N,O-CMC: montmorillonite of 5:1 indicated the disordered intercalated or exfoliated structure in N,O-CMC–montmorillonite nanocomposite. The characteristic IR absorption bands of the symmetric and asymmetric stretching vibrations of $–CH_2$ (2885 cm^{-1}) and $–CH_3$ (2921 cm^{-1}) of intercalated N,O-CMC were observed for the clay composites and the intensity of the bands increased with increasing the molar ratio of N,O-CMC to montmorillonite.

Chitosan intercalated montmorillonite was also prepared by Monvisade and Siriphannon (2009) by slow adding chitosan solution (4 g chitosan in 196 mL of 2% v/v acetic acid) into Na^+-montmorillonite suspension (2.5 g Na^+-montmorillonite in 100 mL water) followed by vigorous stirring of the mixture for 24 h at 60 oC. The parent clay had a basal spacing of 1.23 nm, which increased to ~1.36 nm and ~2.25 nm after intercalation with chitosan. The broad reflection around 2θ of ~6.6° (d_{001}~1.36 nm) was assumed to be due to monolayers of chitosan in montmorillonite whereas d_{001}~2.25 nm was related to intercalation of chitosan bilayers. The IR band at 1594 cm^{-1} attributed to $–NH_2$ group in chitosan shifted to 1520 cm^{-1} in the chitosan-modified montmorillonite, corresponding to deformation vibration of the protonated amine group ($–NH_3^+$) of chitosan. The specific BET surface areas of the parent and modified montmorillonite was 46 and 23 m^2 g^{-1} respectively. The lower surface area of modified clay was attributed to the compact packing of the chitosan molecules in the interlayer space, resulting in pore blocking that inhibited the passage of nitrogen molecules.

The synthesis of chitosan/montmorillonite membranes were reported by Nesic et al. (2012) where the membranes were prepared by mixing solutions of chitosan and montmorillonite in different ratios. The characteristic IR bands needed to confirm complexation between chitosan and montmorillonite shifted to higher frequency. The characteristic bands of montmorillonite appeared at 3742 cm^{-1} (–OH), ~ 3445 cm^{-1} (hydroxyl group from clay and chitosan), 1118–1040 cm^{-1} (stretching vibrations of Si-O) and 523 cm^{-1} (bending vibration of Si- O). Characteristic IR bands of amide groups of chitosan at 1635 cm^{-1} and NH_2^- groups at 1540 cm^{-1} shifted to higher frequencies of 1645 and 1560 cm^{-1} respectively. This indicated that the acetic acid residue was attached to amine group in the chitosan chain. The shift in the bands for amide and amine groups was related to the electrostatic interactions between these groups and the negatively charged sites in the montmorillonite structure confirming complexation between chitosan and clay.

Cellulose acetate/organo-montmorillonite (CA/OMMT) composites were prepared from cellulose acetate (CA) and organo-montmorillonite (OMMT) (Zhou et al., 2012). 0.8 g NaOH was added to montmorillonite suspension (10.0 g clay in 500 mL water) and was stirred. The slurry (Na-MMT) was centrifuged and dried. 4.0 g of Na-MMT was slowly added to 100 ml solution of cetyltrimethylammonium bromide and stirred for 12 h at 60 oC. The solid (OMMT) was washed, dried and 0.12 g OMMT was dispersed in 1.5 mL of acetone with magnetic stirring and treated by ultrasonication for 1 h. A solution of cellulose acetate (0.08 g of Cellulose acetate in 4.7 ml of acetone) was added to the OMMT suspension (0.12 g of OMMT was dispersed in 1.5 ml acetone), the mixture was stirred at room temperature for 24 h, and allowed to evaporate gradually under ambient conditions. A series of composites were made by varying the content of OMMT.

Chandra et al. (2013) used a natural surfactant, rarasaponin (extracted from the flesh of Sapindus rarak fruit) as the modifying agent for bentonite. To a dispersion of natural bentonite in water (0.5 g bentonite in 25 ml water), 5.0 g rarasaponin was added and stirred for 2 h. The mixture was heated in a microwave oven for 5 min at 700 W. The product was washed, dried and pulverized to particle size ~ 100/120 mesh. It was found that the basal spacing of bentonite was expanded from 14.994 to 16.620 Å after modification, possibly due to intercalation of rarasaponin molecules into bentonite interlayer spaces. IR spectral band at 1251.72 cm^{-1} for rarasaponin–bentonite indicated the C=O stretching of deacylated groups from rarasaponin.

Calcium alginate/organophilic montmorillonite was prepared from sodium alginate, vinyl benzyl triethyl ammonium chloride and montmorillonite (Abou Taleb et al., 2012). A 3 wt% sodium alginate solution was added to vinyl benzyl triethyl ammonium- montmorillonite, stirred for 30 min and deoxygenated by degassing for 10 min. The material was irradiated with to 20 kGy gamma rays in a ^{60}Co gamma cell at a dose rate of 2.86 Gy/s. 3 wt% $CaCl_2$ was used in the final feeding solutions of hydrogels to form a network structure of calcium alginate. The material was cooled and the solvent was evaporated to obtain the final product. The montmorillonite XRD band ($2\theta = 11.389^{\circ}$) corresponding to 7.763Å shifted to 2θ of 5.36° corresponding to 15.7Å for the organo-montmorillonite, indicating that intercalation of the surfactant into the galleries of silicate layers. The characteristic band of the clay did not disappear after polymerization with alginate and was still present at 2θ of 5.36°. The intercalation of organo-montmorillonite into Ca/organo-montmorillonite nanocomposites resulted in an increase in basal spacing from 15.7Å to 17Å as the interplaner distance between the faces of their lamellar crystals. The characteristic IR bands of montmorillonite (OH stretching at 3634 cm^{-1}, Si–O stretching at 1043 cm^{-1}, Al–O stretching at 518 cm^{-1} and Si–O bending at 462 cm^{-1}) were observed in the calcium alginate/organo-montmorillonite, confirming the presence of montmorillonite in the organo-montmorillonite and calcium alginate/organo-montmorillonite nanocomposites. The appearance of 800–850 cm^{-1} band confirmed the presence of an aromatic ring of benzyl triethyl ammonium chloride. Electron diffraction experiments showed that each particle was composed of many small crystal nuclei, indicating that the particles had grown in an aggregation model.

The preparation of alginate encapsulated surfactant-modified pillared clay and Ca^{2+} (CTAB-Al-Mont-PILC) was reported by Lezehari et al. (2010). 2.0 to 6.0 g surfactant-modified pillared clay (CTAB-Al-Mont-PILC) was dispersed in 100 ml water. An alginate suspension (1%, w/w; V =100 ml) was added and the mixture was stirred for 2 h. Once the mixture was homogeneous, it was forced through a micropipette tip by a peristaltic pump and the resulting droplets were collected in a stirred reservoir containing 200 ml of 0.1M $CaCl_2$. After 10 hours, the materials were filtered and washed. The IR spectra of the prepared material showed a dual band around 1600 and 1400 cm^{-1} corresponding to carboxyl groups, a band around 1030 cm^{-1} due to CO stretching vibration and a band at 1050 cm^{-1} due to Si-O-Si groups. CHN vibration at 2900 cm^{-1} and (CH_2-CH_3) vibration at 2700 cm^{-1} were also present for surfactant-modified pillared clays. SEM analysis showed changes in surface morphology due to introduction of CTABAl-Mont-PILC.

KSF-montmorillonite was organofunctionalized with protonated 1-dodecylamine by Silva et al. (2012). 200 g KSF-montmorillonite was allowed to react with 100 mL NaCl (1.0 mol L^{-1}) for 12 h at 323 K. The Na-KSF-montmorillonite was separated, washed and dried. A 100 mL solution of dodecylamine (prepared by mixing 4.38 mL of the organic compound with 0.2 mol L^{-1} aqueous HCl) was added to the clay suspension (10.0 g Na-montmorillonite suspended in 50.0 mL of deionized water). The mixture was allowed to react for 48 h at 298 K. The material obtained was dried at 333 K. XRD data showed that the basal spacing of organically modified clay increased from 1.45 to 1.89 nm suggesting the incorporation of organic moieties into the interlayer spacing of KSF-Na^+. The FTIR bands of organo-montmorillonite (1618 and 1506 cm^{-1}) had lower intensity and were attributed to the asymmetric and symmetric bending of the NH_3^+ groups. The bending vibrations of the methylene groups were observed at 1467 cm^{-1}. The band at 1383 cm^{-1} appeared due to C-N group of the organic modifier, verifying the intercalation of surfactant molecules between the silicate layers. The TG of organo-clays exhibited a significant alteration in the thermal degradation. The typical hydrophilic character of KSF-Na^+ was modified after the reaction with the protonated amine, which was evident by the drastic difference between the percentages of the first mass loss over the temperature range of 299–427 K.

In another work, organo-bentonites were synthesized by using hexamethonium bromide (HM) and tetramethyl ammonium bromide (TMA) (Xu and Zhu, 2007). A mixture of 5.0 g bentonite and 100 ml aqueous solution of either hexamethonium bromide or tetramethyl ammonium bromide was stirred for 24 h (25 oC). The organo-bentonites showed IR absorption bands due to C–H groups of pure surfactants in the spectra of the modified bentonite, suggesting intercalation into the interlamellar space of bentonite. The C–H symmetric bending of (N^+)–CH_3 shifted from 1492 to 1489 cm^{-1} (HM) and 1487 cm^{-1} (TMA) after modification. HM molecules could only lie parallel to the silicate planes in a flat-monolayer with two cation heads linked to the siloxane surface in the bentonite interlayer. The surface area of HM-bentonite increased compared to the parent clay, indicating that HM molecules did not form a continuous organic film in the interlayer space due to their small size and were isolated from each other when adhering to the interlamellar surfaces of bentonite. It was suggested that microporous networks formed in the interlayer space between the organic cations. The BET surface area of TMA-bentonite was larger than that of the natural bentonite, suggesting that TMA molecules were arranged in the interlayer space in a way similar to HM molecules. The BET surface area of TMA-bentonite was significantly larger than that of HM-bentonite. Differences in molecular size and structure between the two surfactants might account for this.

Very recently, hexamethylene bis-pyridinium dibromide modified bentonite was prepared by Gu et al. (2014), where, a mixture of 1.0 g Na-bentonite and 50 mL HEMBP solution (0.44 mmol hexamethylene bis-pyridinium dibromide) was shaken for 3 h at 60^0C. The sample was washed, dried and pulverized to pass through a 200 mesh sieve.

The basal spacing of bentonite was expanded from 1.2294 to 1.3627 nm by intercalation of hexamethylene bis-pyridinium dibromide into the interlayer space. hexamethylene bis-pyridinium dibromide- molecules were shown to remain parallel to the siloxane surfaces forming monolayers. The modified bentonite possessed a pair of strong IR bands at 3137 and 3083 cm^{-1}, assigned to the asymmetric and symmetric stretching vibrations of C-H in pyridinium ring. The bands at 2946 and 2876 cm^{-1} appeared due to asymmetric and symmetric stretching vibrations of H-C-H in methylene group.

Adsorption of Dyes on Organo-bentonite

Organo-bentonites have been used in the laboratory for removal of a large number of synthetic dyes from water. Some of these dyes along with their structure are listed in Table 1. The recent applications of organo-bentonites for removal of dyes from water are summarised in Table 2 giving the adsorption capacities of the organo-clays. In most cases, the interactions have followed second order kinetics and Table 3 gives the second order rate coefficient reported in the literature. It may be noted that the experimental conditions used by different research groups have considerable variations and therefore these are also shown in the tables of data.

The adsorption capacity of cetylpyridiniumchloride-bentonite (K_f: 362.07 mg g^{-1}, q_m 416.66 mg g^{-1}) was greater than that of cetyldimethylbenzylammonium chloride-bentonite (K_f: 262.36 mg g^{-1}, q_m 357.14 mg g^{-1}) for Acid Red 151 (Baskaralingam et al., 2006). The organic cations increased the number of surface adsorption sites and made the surface more positive facilitating adsorption of negatively charged dye ions. The cetylpyridiniumammonium-bentonite contained one pyridine ring and a long chain cetyl group. Cetyldimethylbenzylammonium-bentonite had two more methyl groups in addition to one benzene ring and a long chain cetyl group and this might have hindered the dye molecules from reaching the adsorption sites. The second-order kinetics fitted the interactions and the rate coefficient of cetyldimethylbenzyl-ammonium chloride-bentonite was higher compared to that of cetylpyridiniumchloride-bentonite. The clay-dye interactions were maximum at pH 3.0 and minimum at pH 7.0. The electrostatic attraction between the negatively charged SO_3^- anions of the dye molecules and the positively charged adsorption sites was influenced by availability of more protons at lower pH. At higher pH, the surface becomes negative and unfavourable to adsorption of dye anions, but considerable dye removal was achieved at pH ~ 7 to 10 suggesting a chemisorption mechanism.

Table 1. Structure of a few dyes removed by adsorption on organo-bentonites

Name	Structure	Reference
Acid Red 151	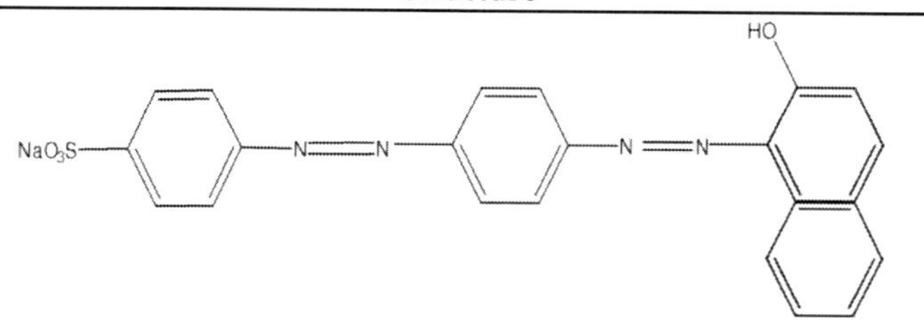	Baskara-lingam et al., 2006
Malachite Green	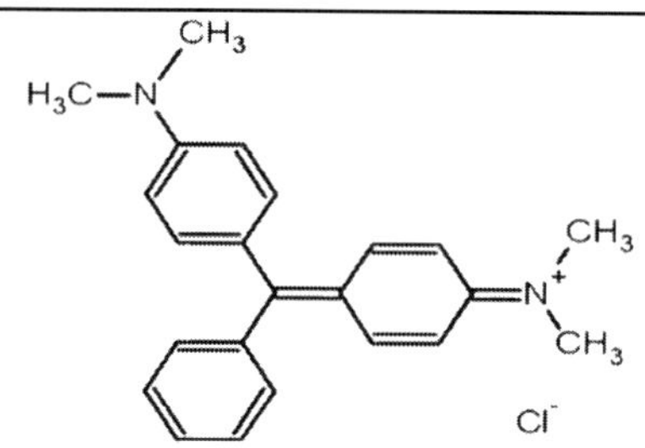	Arellano-Cárdenas et al., 2013
Remazol Brilliant Blue R	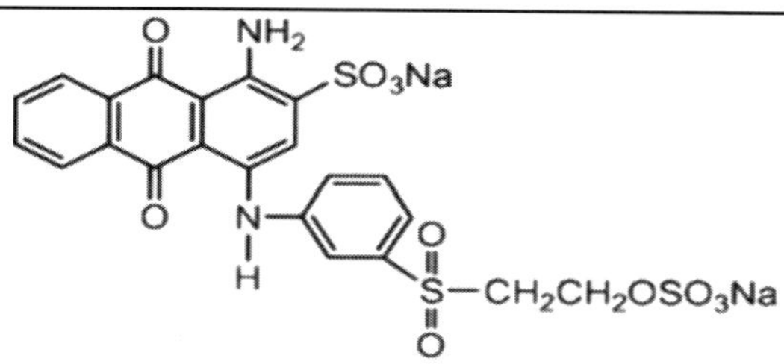	Silva et al., 2012
Reactive Red 141	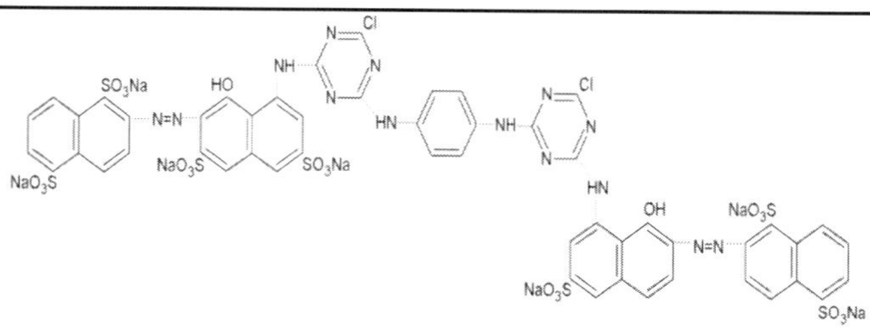	Elemen et al., 2012
Direct Red 2 (Benzopurpurin 4B)	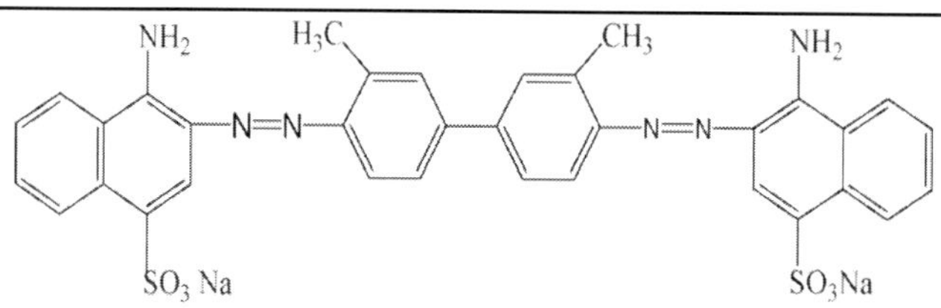	Zohra et al., 2008

Name	Structure	Reference
Acid Scarlet GR	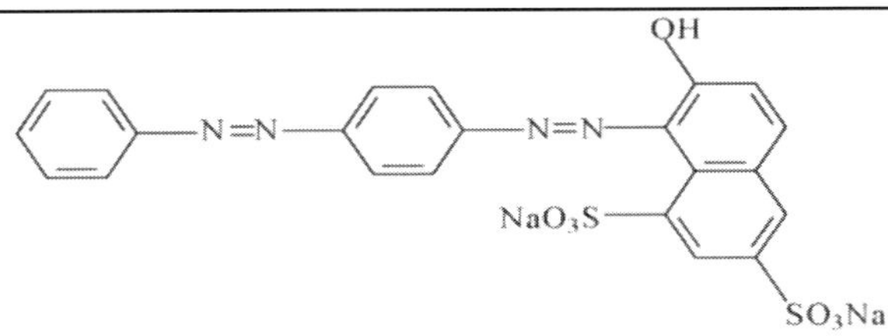	Shen et al., 2009
Acid Turquoise Blue 2G	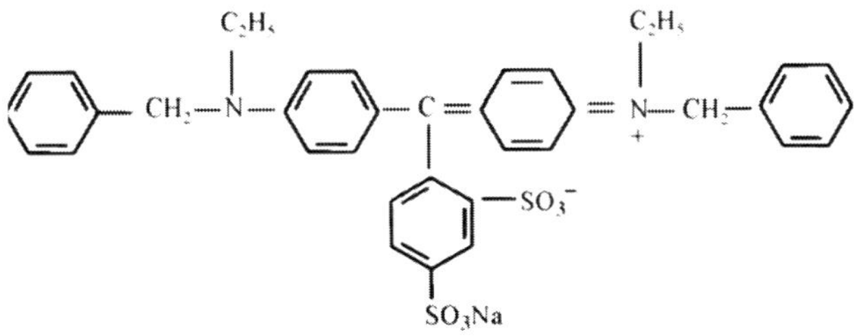	
Indigo Carmine	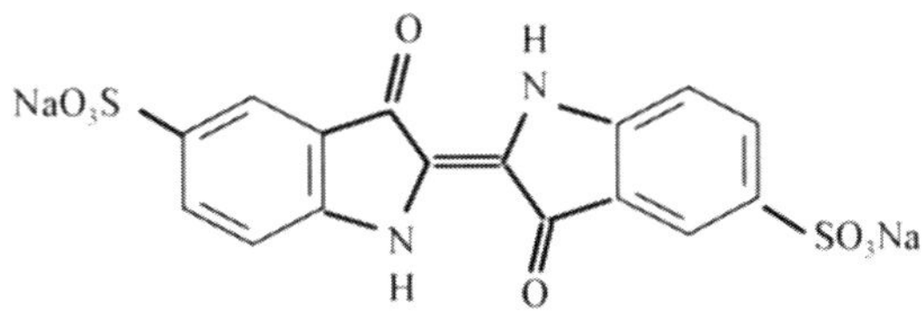	
Reactive Blue 19	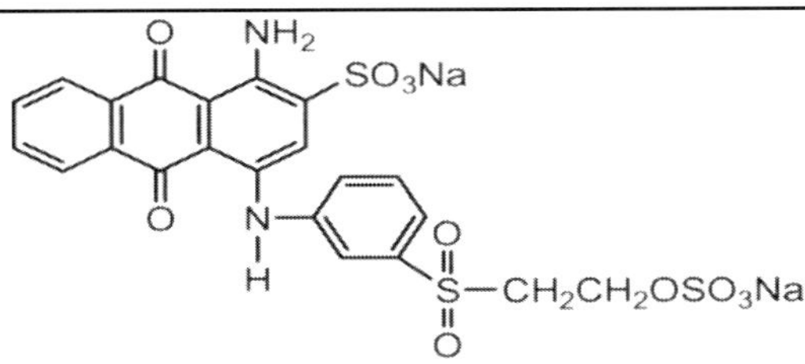	Özcan et al., 2007
Direct Fast Scarlet	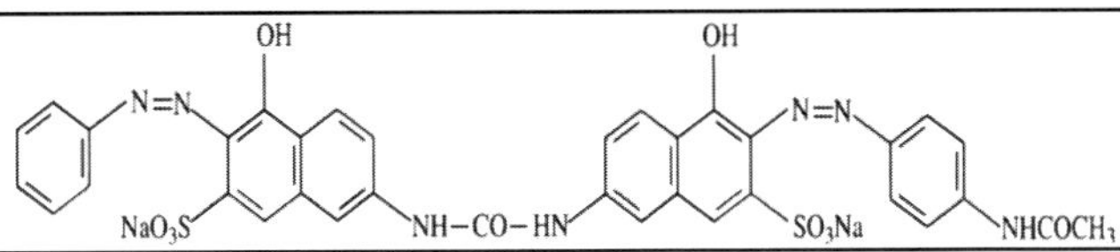	Kang et al., 2009
Eosin Y	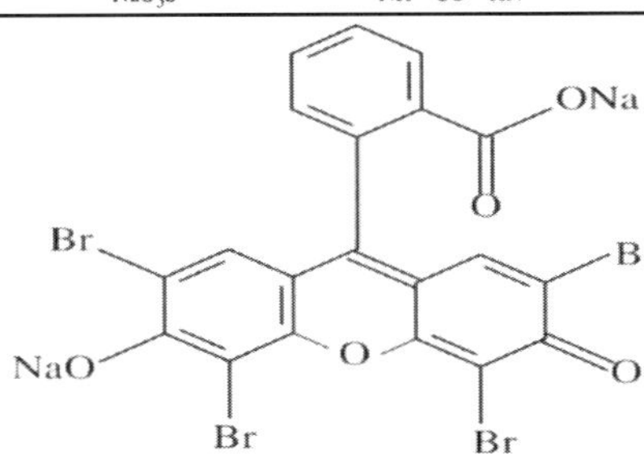	

Table 1. (Continued)

Name	Structure	Reference
Reactive Violet K-3R	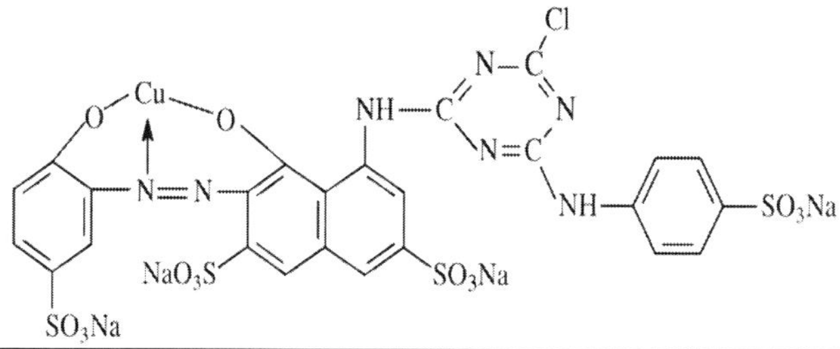	
Acid Dark Blue 2G	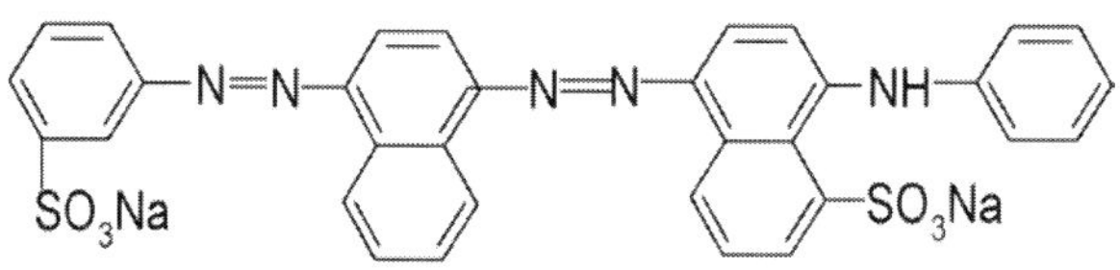	Li et al., 2010
Malachite Green	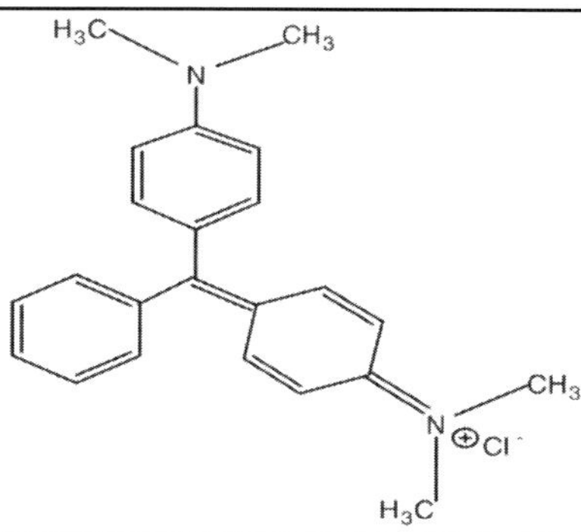	Anirudhan et al., 2009
Methylene Blue	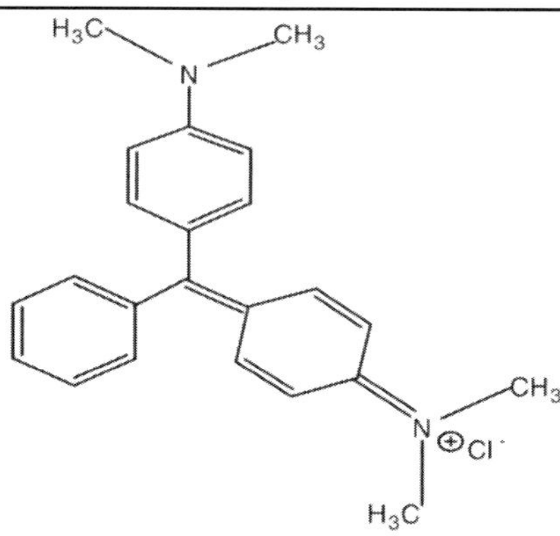	
Crystal Violet	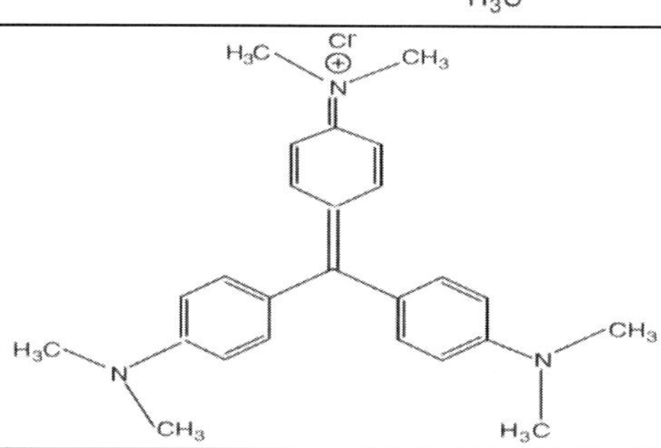	

Name	Structure	Reference
Congo Red	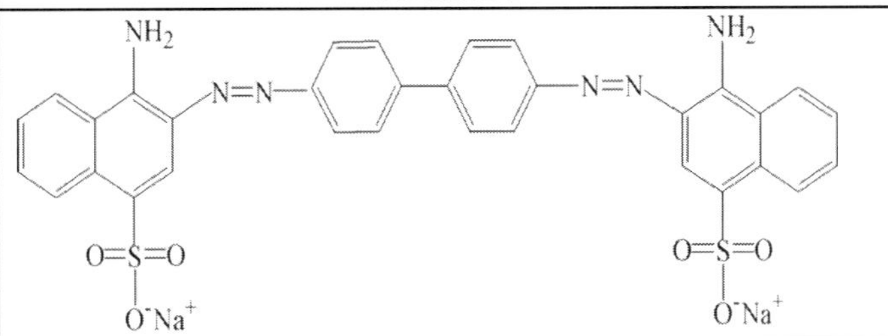	Wang and Wang, 2007
Basic Blue 9	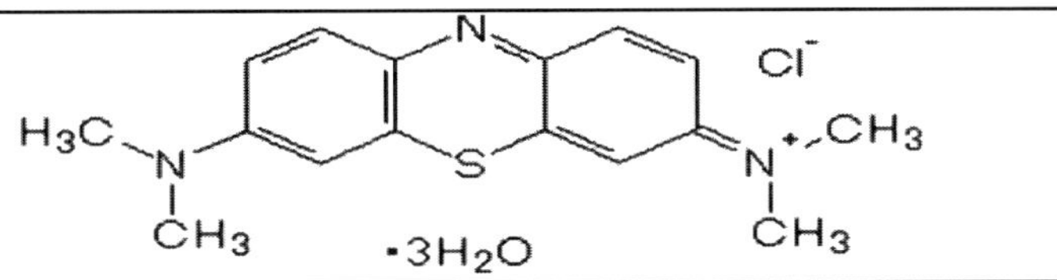	Monvisade and Siriphannon, 2009
Basic Blue 66	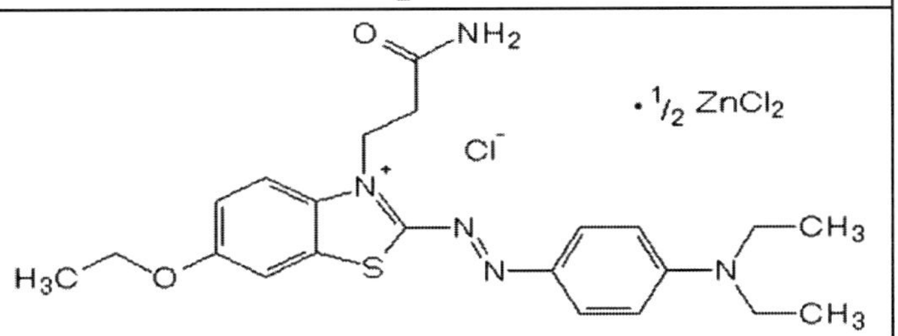	
Basic Yellow 1	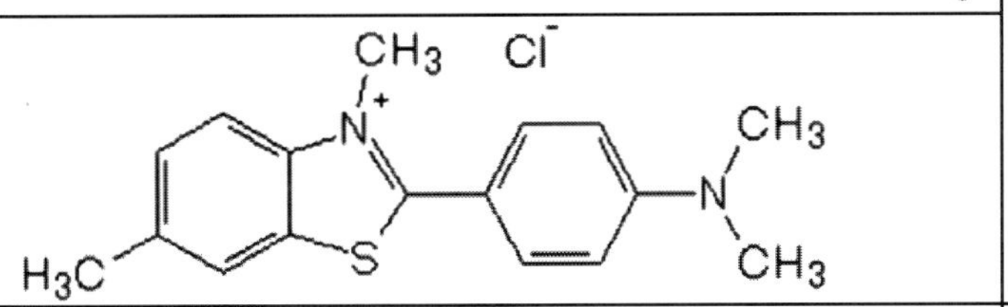	
Bezactiv Orange V-3	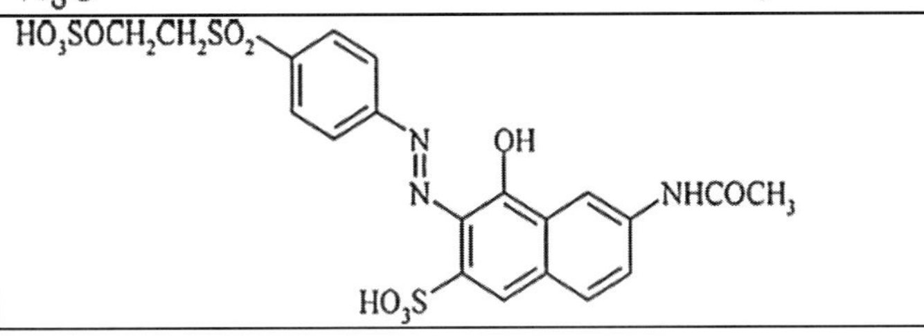	Nesic et al., 2012

The uptake of Orange II on various dual-cationic surfactant-modified organo-bentonites, starting with [HDTMA/HDTMA-bentonite, HDTMA/Myristyltrimethyl-ammonium-bentonite, HDTMA/dodecyltrimethylammonium-bentonite, HDTMA/octyl-trimethylammonium-bentonite] was higher than that of Orange G (Ma et al., 2011) [q_m for Orange II: 144.08 to 249.62 mg g^{-1}, k_f for Orange II: 103.84 to 297.12 mg g^{-1}; q_m for Orange G: 94.72 to 170.46 mg g^{-1}, k_f for Orange II: 42.55 to 167.10 mg g^{-1}]. Orange G molecules ionizing as divalent anions in solution might be exchanged with two Br^- ions leading to decrease in adsorption. Orange II ionizes only as monovalent anions and hence, adsorbs more. The carbon chain length of the surfactant was shown to have influence on adsorption, although the total number of anionic exchange sites remained the same. By decreasing the surfactant carbon chain length, the adsorption capacities of both the dyes decreased.

Table 2. Adsorption capacities of few dyes on organo-clay

Organo-clay	Dye	Experiment conditions		Freundlich capacity		Langmuir capacity		Reference
Cetylpyridinium-bentonite	Acid Red 151	303 K		362.07	mg g^{-1}	416.66	mg g^{-1}	Baskaralingam et al., 2006
Cetyldimethylbenzylammonium chloride-bentonite				262.36		357.14		
Hexadecyltrimethylammonium/ hexadecyltrimethylammonium-bentonite	Orange II	298 K		197.12	mg g^{-1}	249.62	mg g^{-1}	Ma et al., 2011
Hexadecyltrimethylammonium/ Myristyltrimethylammonium-bentonite				175.07		241.07		
Hexadecyltrimethylammonium/ dodecyltrimethylammonium-bentonite				120.80		181.98		
Hexadecyltrimethylammonium/ octyltrimethylammonium-bentonite				103.84		144.08		
Hexadecyltrimethylammonium/ hexadecyltrimethylammonium-bentonite	Orange G	298 K		167.10	mg g^{-1}	170.46	mg g^{-1}	
Hexadecyltrimethylammonium/ Myristyltrimethylammonium-bentonite				91.46		165.09		
Hexadecyltrimethylammonium/ dodecyltrimethylammonium-bentonite				67.46		136.63		
Hexadecyltrimethylammonium/ octyltrimethylammonium-bentonite				42.55		94.72		
Hexadecyltrimethylammonium-Mexican montmorillonite	Malachite Green	298 K		22.49	$mg^{1-1/n}$ g^{-1} $L^{1/n}$	40.48	mg g^{-1}	Arellano-Cárdenas et al., 2013
		313 K		35.37		51.81		
		333 K		27.99		56.82		
Hexadecyltrimethylammonium-bentonite	Reactive Red 141	303 K	Adsorbent: 0.05 gL^{-1}	67.88	$mg^{1-1/n}$ g^{-1} $L^{1/n}$	153.47	mg g^{-1}	Elemen et al., 2012
		313 K		63.03		151.75		
		303 K	Adsorbent: 0.1 gL^{-1}	57.70		104.48		
		313 K		51.88		104.63		
Cetyltrimethylammonium-bentonite	Acid Green 25	303 K	Adsorbent: 0.01-0.06 g L^{-1}, Dye: 1000 mg L^{-1}	-		2.984	mmol g^{-1}	Koswojo et al., 2010
		313 K		-		3.339		
		323 K		-		3.723		
Cetyltrimethylammonium- bentonite	Remazole Brilliant Blue R	303 K	Dye: 50-1000 mg L^{-1}	34.68	L g^{-1}	140.85	mg g^{-1}	Bhatt et al., 2012
		313 K		39.43		153.85		
		323 K		52.12		153.85		

Organo-clay	Dye	Experiment conditions		Freundlich capacity		Langmuir capacity		Reference
Cetyltrimethylammonium-bentonite	Direct Red 2	293 K	30-50 mg L^{-1}	62.01	L g^{-1}	109.89	mg g^{-1}	Zohra et al., 2008
		313 K		88.28		133.33		
		333 K		108.98		153.84		
Octadecyl trimethylammonium-montmorillonite	Acid Red G	303 K		-		3.37×10^{-4}	mol g^{-1}	Tong et al., 2010
		313 K		-		3.24×10^{-4}		
		323 K		-		3.07×10^{-4}		
Dodecyltrimethylammonium-bentonite	Acid Blue 193	293 K		127.10	L mg^{-1}	1635.4	mg g^{-1}	Özcan et al., 2004
		303 K		54.70		2011.0		
		313 K		46.90		1732.0		
		323 K		3.05		4153.0		
Dodecyltrimethylammonium-bentonite	Reactive Blue 19	293 K	Dye: (2.0-4.5) x 10^{-4} mol L^{-3}	1.56×10^{-3}	L g^{-1}	3.30×10^{-4}	mol g^{-1}	Özcan et al., 2007
		303 K		1.51×10^{-3}		3.15×10^{-4}		
		313 K		1.46×10^{-3}		2.98×10^{-4}		
Montmorillonite	Methylene Blue		Dye: 20-200 mg L^{-1}			322.4	mg g^{-1}	Wibulswas, 2004
100% CEC-TMA						172.4		
200% CEC-TMA						217.4		
50% CEC-HDTMA						243.9		
100% CEC-HDTMA						227.3		
200% CEC-HDTMA						13.7		
50% CEC-TDMA						163.9		
100% CEC-TDMA						178.6		
100% CEC-TDMA						208.3		
Polydiallydimethylammonium-bentonite	Acid Turquoise Blue 2G	298 K		-		176.3±9.1	µmol L^{-1}	Shen et al., 2009
	Acid Scarlet GR			-		228.7 ±12.4		
	Indigo Carmine			-		149.2 ±8.3		
Polydiallydimethylammonium-bentonite	Acid Turquoise Blue 2G	298 K		-		176.3 ±9.1	µmol L^{-1}	Shen et al., 2009
	Acid Scarlet GR			-		228.7 ±12.4		
	Indigo Carmine			-		149.2 ±8.3		
Poly-epichlorohydrin-dimethylamine-bentonite	Disperse Blue SBL	298 K		0.672	$mg^{1-1/n} g^{-1} L^{1/n}$	9.52	mg g^{-1}	Li et al., 2010
	Vat Scarlet R			0.272		16.43		
	Reactive Violet K-3R			19.612		44.64		
	Acid Dark Blue 2G			12.834		40.77		

Table 2. (Continued)

Organo-clay	Dye	Experiment conditions		Freundlich capacity		Langmuir capacity		Reference
Poly-epicholorohydrin- dimethylamine-bentonite	Direct Fast Scarlet	298±2 K		-		148.6 ±2.4	mg g^{-1}	Kang et al., 2009
	Reactive Violet K-3R			-		106.7 ±1.9		
	Eosin Y			-		88.4 ±2.1		
Poly-epicholorohydrin-dimethylamine-bentonite	Acid Scarlet GR	293 K		15.23	$mg^{1-1/n} g^{-1} L^{1/n}$	42.19	mg g^{-1}	Li et al., 2010
		303 K		21.32		45.54		
		313 K		25.90		51.79		
	Acid Dark Blue 2G	293 K		20.58	$mg^{1-1/n} g^{-1} L^{1/n}$	34.34	mg g^{-1}	
		303 K		23.27		35.93		
		313 K		25.14		40.98		
Humic acid immobilized polyacrylamide bentonite composites with amine functionality	Malachite green	303 K	Dye: 20-200 µmol L^{-1}, pH: 6.0	142.1	Lg^{-1}	656.6	µmol g^{-1}	Anirudhan et al., 2009
	Methylene Blue			89.7		648.4		
	Crystal Violet			71.8		510.4		
2-aminomethylpyridine immobilized-smectite	Sumifix Brilliant Orange 3R	298±1 K	Dye: 0.125-3.5 mmol L^{-1}			2.12 ±0.01	mmol g^{-1}	Guerra et al., 2011
Para-bis(triphenyl phosphonium methylene)-benzene- montmo-rillonite	Telon Red	294 K	Dye: 10 - 500 mg L^{-1}	1.05		84.0	mg g^{-1}	Makhoukhi et al., 2010
	Telon Blue			1.47		62.5		
	Telon Orange			1.04		110.0		
Ortho-bis(triphenyl phosphonium methylene)-benzene- montmo-rillonite	Telon Red			1.10		71.4	mg g^{-1}	
	Telon Blue			1.45		41.7		
	Telon Orange			0.80		76.9		
Meta-bis(triphenyl phosphonium methylene)-benzene- montmo-rillonite	Telon Red			1.02		78.7	mg g^{-1}	
	Telon Blue			1.10		47.5		
	Telon Orange			0.97		100.0		
Organofunctionalized (with protonated 1-dodecylamine)-montmorillonite	Remazol Brilliant Blue R	298 K		7.38	$mg^{1-1/n} g^{-1} L^{1/n}$	30.76	mg g^{-1}	Silva et al., 2012
		303 K		6.42		34.24		
		313 K		9.41		32.05		
		323 K		15.01		31.74		
1,6-diamino hexane-bentonite	Reactive Blue 19	293 K	Adsorbent: 0.01-0.04 g L^{-1}	6.31×10^{-3}	L g^{-1}	2.15×10^{-4}	mol g^{-1}	Gök et al., 2010
Chitosan-bentonite	Acid Scarlet	298 K		12.257		102.04	mg g^{-1}	Guo et al., 2012
Chitosan/hexadecyl trimethyl ammonium-bentonite				37.483		175.44		

Organo-clay	Dye	Experiment conditions		Freundlich capacity		Langmuir capacity		Reference
Chitosan/montmorillonite nanocomposite	Congo Red	303 K	-	-		54.52	mg g^{-1}	Wang and Wang, 2007
N,O-carboxymethyl-chitosan/montmorillonite	Congo Red	303 K	-	-		74.14	mg g^{-1}	Wang and Wang, 2008
zChitosan intercalated montmorillonite	Basic Blue 9					48.90	mg g^{-1}	Monvisade and Siriphannon, 2009
	Basic Blue 66					49.20		
	Basic Yellow 1					45.90		
Chitosan/Montmorillonite 10%	Bezactiv Orange V-3R	293 K	Dye 20-70 mg L^{-1}	34.10	mg g^{-1}	106.80	mg g^{-1}	Nesic et al., 2012
Chitosan/Montmorillonite 20%				37.80		128.70		
Chitosan/Montmorillonite 30%				21.10		277.70		
Chitosan/Montmorillonite 40%				30.80		322.60		
Chitosan/Montmorillonite 50%				20.40		740.70		
Cellulose acetate/organo-montmorillonite composite	Acid Scarlet G	293 K		$10.47 x 10^3$	L g^{-1}	$2.97 x 10^4$	mol g^{-1}	Zhou et al., 2012
		313 K		$6.27 x 10^3$		$2.88 x 10^4$		
		333 K		$8.96 x 10^3$		$3.04 x 10^4$		
Chitosan-*g*-poly(acrylic acid)/ montmorillonite nanocomposite	Methylene Blue	303 K	pH: 6.5	$1.81 x 10^3$	$mg^{1-1/n} g^{-1} L^{1/n}$	1859.0	mg g^{-1}	Wang et al., 2008
Rarasaponin-bentonite	Evans Blue	333 K		0.62	$mmol^{1-1/n} g^{-1} L^{1/n}$	0.53	mmol g^{-1}	Chandra et al., 2013
Rarasaponin-bentonite	Methylene Blue	303 K		9.30	$mg^{1-1/n} g^{-1} L^{1/n}$	231.0	mg g^{-1}	Kurniawan et al., 2011
		318 K		28.11		247.0		
		333 K		68.72		256.0		
Alginate encapsulated surfactant-modified pillared bentonite	Safranine	295 K	Dye 285- 4000 μmol L^{-1}	141.0	$\mu mol^{1-1/n} L^{1/n} g^{-1}$	705.0	μmol g^{-1}	Lezehari et al., 2010

Table 3. Second-order rate coefficient (k_2) for adsorption of a few dyes on organo-clays

Organo-clay	Dye	Experimental conditions		k_2		Reference
Hexadecyltrimethylammonium-Mexican montmorillonite	Malachite Green	298	Dye: 60 mg L^{-1}, pH: 6.0	0.006	g mg^{-1} min^{-1}	Arellano-Cárdenas et al., 2013
		313		0.027		
		333		0.597		
Hexadecyltrimethylammonium-bentonite	Reactive Red 141	303 K	Adsorbent: 0.05 g L^{-1}	0.0024	L^3 mol^{-1} s^{-1}	Elemen et al., 2012
		313 K		0.0023		
		303 K	Adsorbent: 0.1 g L^{-1}	0.0491		
		313 K		0.0424		
Cetylpyridinium-bentonite	Acid Red 151	303 K	Dye: 60 mg L^{-1}	1.138 x 10^{-3}	g mg^{-1} min^{-1}	Baskaralingam et al., 2006
			Dye: 80 mg L	1.366 x 10^{-3}		
			Dye: 90 mg L	1.415 x 10^{-3}		
Cetyldimethylbenzylammonium-bentonite	Acid Red 151	303 K	Dye: 70 mg L^{-1}	5.264 x 10^{-4}		
			Dye: 80 mg L	6.062 x 10^{-4}		
			Dye: 100 mg L	3.439 x 10^{-4}		
Cetyltrimethylammonium-bentonite	Acid Green 25	303 K	Adsorbent: 0.01-0.06 g L^{-1}, Dye: 1000 mg L^{-1}	0.036	g $mmol^{-1}$ min^{-1}	Koswojo et al., 2010
		313 K		0.029		
		323 K		0.023		
Cetyltrimethylammonium-bentonite	Remazole Brilliant Blue R	303 K	Adsorbent: 0.2 gL^{-1}, Dye: ~488 mg L^{-1}	0.73 x 10^3	g mg^{-1} min^{-1}	Bhatt et al., 2012
Cetyltrimethylammonium-bentonite	Direct Red 2	293 K	Dye: 120 mg L^{-1}, pH: 6.3	10.13 x 10^{-3}		Zohra et al., 2008
		313 K		8.35 x 10^{-3}		
		333 K		4.83 x 10^{-3}		
		293 K	Dye: 45 mg L; pH: 6.3	2.95 x 10^{-3}		
			Dye: 75 mg L; pH: 6.3	6.09 x 10^{-3}		
			Dye: 120 mg L; pH: 6.3	10.13 x 10^{-3}		

Organo-clay	Dye	Experimental conditions		k$_2$		Reference
Dodecyltrimethylammonium-bentonite	Acid Blue 193	293 K		6.32×10^{-4}	g mg^{-1} min^{-1}	Özcan et al., 2004
		303 K		7.80×10^{-4}		
		313 K		1.16×10^{-3}		
		323 K		$1.47 \times 10^{-}$		
		333 K		1.83×10^{-3}		
Dodecyltrimethylammonium-bentonite	Reactive Blue 19	293 K	Dye: (2.0-4.5) $\times 10^{-4}$ mol L^{-3}	3.60×10^{3}	g mol^{-1} min^{-1}	Özcan et al., 2007
		303 K		5.65×10^{3}		
		313 K		1.15×10^{3}		
Octadecyltrimethylammonium-montmorillonite	Acid Red G	303K	Dye: 100 mg L^{-1}	0.571	g mg^{-1} min^{-1}	Tong et al., 2010
			Dye: 150 mg L^{-1}	0.136		
			Dye: 200 mg L^{-1}	0.040		
			Dye: 250 mg L^{-1}	0.015		
			Dye: 300 mg L^{-1}	0.005		
		313 K	Dye: 100 mg L^{-1}	0.311		
			Dye: 150 mg L^{-1}	0.138		
			Dye: 200 mg L^{-1}	0.014		
			Dye: 250 mg L^{-1}	0.004		
			Dye: 300 mg L^{-1}	0.001		
		323 K	Dye: 100 mg L^{-1}	1.485		
			Dye: 150 mg L^{-1}	0.150		
			Dye: 200 mg L^{-1}	0.027		
			Dye: 250 mg L^{-1}	0.023		
			Dye: 300 mg L^{-1}	0.001		
Polydiallydimethylammonium-bentonite	Acid Scarlet GR	298 K	Adsorbent: 0.5 g L^{-1}	239.0	g mol^{-1} min^{-1}	Shen et al., 2009
			Adsorbent: 1.0 g L^{-1}	896.0		
			Adsorbent: 2.0 g L^{-1}	6.17×10^{3}		
			Adsorbent: 3.0 g L^{-1}	1.49×10^{3}		

Table 3. (Continued)

Organo-clay	Dye	Experimental conditions			k_2		Reference
	Indigo Carmine	298 K	Adsorbent: 0.5 g L^{-1}		379		
			Adsorbent: 1.0 g L^{-1}		786.0		
			Adsorbent: 2.0 g L^{-1}		4.62 x 10^3		
			Adsorbent: 3.0 g L^{-1}		1.02 x 10^3		
	Acid Turquoise Blue 2G	298 K	Adsorbent: 0.5 g L^{-1}		172.0		
			Adsorbent: 1.0 g L^{-1}		421.0		
			Adsorbent: 2.0 g L^{-1}		2.07 x 10^3		
			Adsorbent: 3.0 g L^{-1}		6.46 x 10^3		
Poly-epichlorohydrin-dimethylamine-bentonite	Acid Scarlet GR	293 K	Dye: 50 mg L^{-1}, pH: 7.5		43.1 x10^{-3}	g mg^{-1} min^{-1}	Li et al., 2011
		303 K			20.3 x10^{-3}		
		313 K			17.3 x10^{-3}		
	Acid Dark Blue 2G	293 K	Dye: 50 mg L^{-1}, pH: 7.9		6.65 x10^{-3}		
		303 K			4.60 x10^{-3}		
		313 K			3.37 x10^{-3}		
Poly-epichlorohydrin-dimethylamine-bentonite	Disperse Blue SBL	298 K			0.0885	g mg^{-1} min^{-1}	Li et al., 2010
	Vat Scarlet R				0.0378		
	Reactive Violet K-3R				0.0034		
	Acid Dark Blue 2G				0.0044		
Poly-epicholorohydrin- dimethylamine-bentonite	Direct Fast Scarlet	298±2 K	Dye 120 mg L^{-1}	Clay 0.5 g L^{-1}	(87.6±5.5) x10^{-5}	g mg^{-1} min^{-1}	Kang et al., 2009
				Clay 1.0 g L^{-1}	(126±7) x10^{-5}		
				Clay 1.5 g L^{-1}	(313±21) x10^{-5}		
				Clay 2.0 g L^{-1}	(662±4.7) x10^{-5}		

Organo-clay	Dye	Experimental conditions			k_2		Reference
	Reactive Violet K-3R	298±2 K	Dye 120 mg L^{-1}	Clay 0.5 g L^{-1}	(71.4±5.8) $x10^{-5}$		
				Clay 1.0 g L^{-1}	(86.7±6.3) $x10^{-5}$		
				Clay 1.5 g L^{-1}	(135±6) $x10^{-5}$		
				Clay 2.0 g L^{-1}	(267±12) $x10^{-5}$		
	Eosin Y	298±2 K	Dye: 120 mg L^{-1}	Clay 0.5 gL^{-1}	(184±12) $x10^{-5}$		
				Clay 1.0 gL^{-1}	(253±12) $x10^{-5}$		
				Clay 1.5 gL^{-1}	(351±15) $x10^{-5}$		
				Clay 2.0 gL^{-1}	(516±19) $x10^{-5}$		
Humic acid immobilized polyacrylamide bentonite composites with amine functionality	Malachite green	303 K	pH: 6.0	Dye 400 µmol L^{-1}	2.97 x 10^{-2}	mg g^{-1} s^{-1}	Anirudhan et al., 2009
				Dye 600 µmol L^{-1}	2.25 x 10^{-2}		
				Dye 800 µmol L^{-1}	1.82 x 10^{-2}		
				Dye 1000 µmol L^{-1}	1.43 x 10^{-2}		
	Methylene Blue			Dye 400 µmol L^{-1}	2.95 x 10^{-2}	mg g^{-1} s^{-1}	
				Dye 600 µmol L^{-1}	2.16 x 10^{-2}		
				Dye 800 µmol L^{-1}	1.68 x 10^{-2}		
				Dye 1000 µmol L^{-1}	1.32 x 10^{-2}		

Table 3. (Continued)

Organo-clay	Dye	Experimental conditions		k_2		Reference
	Crystal Violet		Dye 400 μmol L^{-1}	2.64×10^{-2}	mg g^{-1} s^{-1}	
			Dye 600 μmol L^{-1}	2.20×10^{-2}		
			Dye 800 μmol L^{-1}	1.60×10^{-2}		
			Dye 1000 μmo lL^{-1}	1.25×10^{-2}		
Organofunctionalized (with protonated 1-dodecylamine)-montmorillonite	Remazol Brilliant Blue R	298	Dye 100 mg L^{-1}	0.0043	g mg^{-1} min^{-1}	Silva et al., 2012
		303		0.0033		
		313		0.0034		
		323		0.0035		
2-aminomethylpyridine immobilized-smectite	Sumifix Brilliant Orange 3R	298 K	Dye 0.125-3.5 mmol L^{-1}	91.14±0.05	mmol g^{-1} min^{-1}	Guerra et al., 2011
		303 K		95.15±0.04		
		308 K		98.13±0.08		
		313 K		99.01±0.05		
1,6-diamino hexane-bentonite	Reactive Blue 19	293 K	Dye 1.00×10^{-4} mol L^{-1}	7.40×10^{4}	g mol^{-1} min^{-1}	Gök et al., 2010
			Dye 1.50×10^{-4} mol L^{-1}	7.14×10^{4}		
			Dye 2.00×10^{-4} mol L^{-1}	5.69×10^{4}		
			Dye 1.50×10^{-4} mol L^{-1}	4.20×10^{4}		
			Dye 3.00×10^{-4} mol L^{-1}	3.54×10^{4}		
			Dye 3.50×10^{-4} mol L^{-1}	2.67×10^{4}		
		303 K	Dye 2.20×10^{-4} mol L^{-1}	5.88×10^{4}		
		313 K		6.77×10^{4}		

Organo-clay	Dye	Experimental conditions		k_2		Reference
Chitosan/Montmorillonite 10%	Bezactiv Orange V-3R		Dye 30 mg L^{-1}	2.55×10^{-3}	g mg^{-1} h^{-1}	Nesic et al., 2012
Chitosan/Montmorillonite 20%				4.49×10^{-3}		
Chitosan/Montmorillonite 30%				3.53×10^{-3}		
Chitosan/Montmorillonite 40%				3.27×10^{-3}		
Chitosan/Montmorillonite 50%				3.31×10^{-3}		
Chitosan/Montmorillonite 10%			Dye 50 mg L^{-1}	1.83×10^{-3}		
Chitosan/Montmorillonite 20%				1.86×10^{-3}		
Chitosan/Montmorillonite 30%				1.19×10^{-3}		
Chitosan/Montmorillonite 40%				5.40×10^{-4}		
Chitosan/Montmorillonite 50%				1.17×10^{-4}		
Chitosan/Montmorillonite 10%			Dye 80 mg L^{-1}	2.67×10^{-3}		
Chitosan/Montmorillonite 20%				1.20×10^{-3}		
Chitosan/Montmorillonite 30%				5.50×10^{-4}		
Chitosan/Montmorillonite 40%				4.60×10^{-4}		
Chitosan/Montmorillonite 50%				2.60×10^{-4}		
Rarasaponin-bentonite	Evans Blue	303 K		0.387	g mol^{-1} min^{-1}	Chandra et al., 2013
		318 K		0.338		
		333 K		0.268		
Rarasaponin-bentonite	Methylene Blue		Dye: 300 mgL^{-1}	0.031	mg g^{-1} min^{-1}	Kurniawan et al., 2011
			Dye: 500 mgL^{-1}	0.022		
			Dye: 700 mgL^{-1}	0.017		
Alginate encapsulated surfactant-modified pillared bentonite	Safranine	295 K	Dye: 285μmol L^{-1}	0.034	μmol h	Lezehari et al., 2010

Langmuir adsorption capacities of hexadecyltrimethylammonium modified-Mexican montmorillonite for malachite green, increased from 40.48 to 56.82 mg g^{-1} with temperature increasing from 298 to 333 K (Arellano-Cárdenas et al., 2013). The clay-dye interactions followed second order kinetics and chemisorption interaction were shown to control the rate-controlling step. The uptake of the dye increased slightly from 95.0 % at pH 3.0 to 99.6 % in the pH interval of 4.0 – 10.0. The dye molecules got protonated in the acidic medium and deprotonated at higher pHs. Further the surface groups of the clay adsorbent were protonated at pH < 8.0. The hydrophobic and chemical interactions could be the key forces responsible for adsorption.

The adsorption capacity of organofunctionalized (with protonated 1-dodecylamine)-montmorillonite for Remazol brilliant blue R was influenced by solution temperature (Silva et al., 2012). The q_m values increased from 30.76 to 34.24 mg g^{-1} by increasing the temperature from 298 to 303 K and again decreased to 31.74 mg g^{-1} at 323 K. At the higher temperatures (303 K), the mobility of the dye molecules improved facilitating their penetration and accessibility to active sites in the porous structure of the solid, which is cited as the reason for enhancing adsorption. However, further rise in temperature decreased adsorption. The adsorption interactions followed second-order kinetic model with a rate coefficient of 0.0043 g $mg^{-1}min^{-1}$ at 303 K.

Hexadecyltrimethylammonium-bentonite adsorbed Reactive Red 141 much more at a lower loading (0.05 g L^{-1}) than a higher one (0.1 g L^{-1}) (Elemen et al., 2012). No significant difference was observed in the adsorption capacities at two different temperatures, 303 and 313 K. The dye molecules were shown to be held on the modified clay through hydrophobic interactions. The clay-dye interactions were predicted by applying a three-layered neural network with 5 neurons in the hidden layer and using a backpropagation algorithm. The analysis of the relationship between the predicted results of the designed model and the experimental data revealed the values of the determination coefficient and the mean square error to be 0.978 and 0.027364, respectively.

Koswojo et al. (2010) reported the validity of the Langmuir adsorption model in adsorption of Acid Green 25 on cetyltrimethylammonium-bentonite surface. Higher temperature favoured adsorption and the adsorption capacity increased from 2.984 to 3.723 mmol g^{-1} in the temperature range of 303 to 323 K, indicating endothermic interactions. Application of the second-order kinetics yielded a rate coefficient of 0.036 to 0.023 g $mmol^{-1}$ min^{-1} (303 to 323 K). The trend suggested a chemisorption reaction as the predominant rate-controlling step.

Adsorption of Remazole Brilliant Blue R on cetyltrimethylammoniumbromide- bentonite yielded Langmuir capacities in the range of 140.85 to 153.85 mg g^{-1} in the temperature range of 303 to 333 K (Bhatt et al., 2012). The surfactant cations were intercalated between the layers of the bentonite forming a hydrophobic lamellar type porous structure. At higher equilibrium concentration, an increase in dye adsorption was observed with temperature. This might be due to a higher partitioning rate of dye molecules in organophilic clays. The second-order rate coefficient for clay-dye interactions was 0.73 x 10^3 g $mg^{-1}min^{-1}$.

Langmuir capacities of cetyltrimethylammonium-bentonite for adsorption of the dye, Direct Red 2 (Benzopurpurin 4B) was enhanced from 109.89 to 153.84 mg g^{-1} in the temperature range of 293 to 333 K (Zohra et al., 2008), indicating the adsorption process to be endothermic. The adsorptive forces between the active sites of organophilic clay and dye species were stronger at a higher temperature, partially due to the weakening of the forces between the adjacent molecules of the adsorbed phase. The authors suggested that the good adsorption capacity of the modified clay was due to the alkyl chains in the interlamellar spaces and also due to the particular structure of the dye. The two sulfonic groups ($-SO_3$) and the anthraquinone ring in the molecular structure might play some key role in the process. The adsorptive interactions were suggested to have followed a pseudo-second order model, based on the assumption that the rate-limiting step might be chemical adsorption involving valence forces through sharing or exchange electrons between the adsorbent and the adsorbate.

The interactions of Acid Red G onto octadecyl trimethylammonium-montmorillonite yielded Langmuir capacities that decreased from 3.37×10^{-4} to 3.065×10^{-4} mol g^{-1} by increasing the temperature from 303 to 323 K (Tong et al., 2010).

Shen et al. (2009 studied adsorption of Acid Scarlet GR, Acid Turquoise Blue 2G and Indigo Carmine on polydiallydimethylammonium-bentonite in single, binary and ternary dye solutions. In single dye solutions, the adsorption isotherms of the three dyes were well described by the Langmuir adsorption equation, where the related adsorption capacities were in the order of Acid Scarlet GR > Acid Turquoise Blue 2G > Indigo Carmine. In adsorption from binary and ternary dye solutions, the removal efficiency was higher for the dye that had a greater equilibrium adsorption constant, especially in the insufficient dosage region. The cationic polyelectrolyte template that was used as modifier of the bentonite appreciated electrostatic attraction between the negatively charged SO_3^- anionic groups of the dye and quaternary ammonium groups in the layers of the clays.

Özcan et al. (2004) compared the adsorption of Acid Blue 193 on dodecyltrimethylammonium-bentonite and Na-bentonite and found that the latter was a less effective adsorbent for the dye, a hydrophobic organic compound, from aqueous solution, due to electrically charged and hydrophilic characteristics of the clay surface. The modification with a large organic cation, namely, dodecyltrimethylammonium- significantly improved its capability for accumulating hydrophobic contaminants from water. Thus, the Freundlich capacity increased from 4.51 to 127.1 L mg^{-1} at 293 K after the modification of the clay. The high adsorption capacity was due to the strong electrostatic interactions between the $-N^+(CH_3)_3$ of dodecyltrimethylammonium–bentonite and dye anions. On the contrary, with fewer adsorption sites, Na–bentonite possessed low adsorption capacity due to the lack of electrostatic interactions with the dye anions. The second-order rate coefficients of the adsorption systems increased steadily from 6.32×10^{-4} to 1.83×10^{-3} g $mg^{-1}min^{-1}$ with increase in temperature (293 to 333 K). However, the equilibrium dye adsorption capacity decreased from 740.50 to 366.20 mg g^{-1}. The interactions were shown to be of physisorption type.

The same group of authors used dodecyltrimethylammonium-bentonite as adsorbent for Reactive Blue 19 from aqueous solution (Özcan et al., 2007). The strong electrostatic interactions between the adsorption sites of clay and dye anion might be responsible for the high adsorption capacity (3.30×10^{-4} mol g^{-1}) of organo-bentonite in acidic solutions (~pH 1.5). The Langmuir capacities decreased from 3.30×10^{-4} to 2.98×10^{-4} mol g^{-1} as the solution temperature enhanced from 203 to 313 K. It was assumed that higher temperature increased the total energy of the adsorbate molecules along with their escaping tendency, thus, the adsorption of the adsorbate species decreased. The clay-dye interactions were favoured by second-order kinetics with a rate coefficient of 3.60×10^3 g $mol^{-1}min^{-1}$ at 293 K.

Organo-modified montmorillonites [modified with tetramethylammonium- (TMA), tetradecyltrimethylammonium- (TDMA), hexadecyltrimethylammonium- (HDTMA) and benzyldimethylhexadecylammonium- (BDHDMA)] were used for removal of methylene blue from aqueous solution (Wibulswas, 2004). It was observed that methylene blue was removed more effectively by montmorillonite than by the organo-clays and it was not adsorbed at all by BDHDMA-clay. This was because methylene blue is not only soluble in water, indicating its hydrophilic nature, but it also ionizes in solution yielding cations which can undergo cation exchange reactions with montmorillonite. The increased surfactant load on the clay lowered methylene blue uptake. The surfactant occupied the cation-exchange sites of the precursor clay, making it more difficult for methylene blue cations to be adsorbed, so that equilibrium was reached with lower concentration of the dye.

The kinetics of adsorption of two acid dyes, Acid Scarlet GR and Acid Dark Blue 2G, onto poly-epichlorohydrin-dimethylamine modified bentonite was reported by Li et al. (2011). The interactions followed second-order model with a high degree of correlation (r~0.99) whereas the intraparticle diffusion model presented two steps with two different diffusion rate constants indicating that the intraparticle diffusion may be involved in the adsorption process during the first 15 min. The second order rate coefficient for Acid Scarlet GR and Acid Dark Blue 2G varied from 43.1 $x10^{-3}$ to 17.3 $x10^{-3}$ g $mg^{-1}min^{-1}$ and 6.65 $x10^{-3}$ to 3.37 $x10^{-3}$ g $mg^{-1}min^{-1}$, respectively.

Langmuir adsorption capacities of poly-epicholorohydrin-dimethylamine-bentonite for Acid Scarlet GR and Acid Dark Blue 2G were found to increase with solution temperature indicating endothermic interactions (Li et al., 2010). It was suggested that electrostatic attraction between the strongly negative dye ions and the organo-clay might be responsible for adsorption. Poly-epichlorohydrin-dimethylamine-bentonite was also used for adsorption of a few non-ionic dyes (Disperse Blue SBL and Vat Scarlet R) and anionic dyes (Reactive Violet K-3R and Acid Dark Blue 2G) (Li et al., 2010). The adsorbent possessed highest adsorption capacity for anionic dyes compared to the non-ionic dyes. The non-ionic dyes existed as colloidal particles in solution and the electrostatic attraction between the colloidal particles and the modified clay surface was the key factor in adsorption. For the anionic dyes, besides electrostatic interactions between the dyes and the adsorbent, $-SO_3^-$ groups in the dye structures might react with the amidocyanogens of EPI-DMA. Thus, the formation of $-NH^{2+}SO_3^-$ or $=NH^+SO_3^-$ between anionic dye molecules and modified bentonite along with the particles decreased the second order rate coefficient.

Poly(epicholorohydrin dimethylamine)-bentonite was also used by Kang et al. (2009) for adsorptive accumulation of three anionic dyes, namely Direct Fast Scarlet, Eosin Y and Reactive Violet K-3R in single, binary and ternary systems. The authors observed that at low dye concentration, the isotherm curves rose steeply indicating complete adsorption with a plateau indicating the maximum adsorption capacity. Langmuir capacity was maximum for Direct Fast Scarlet followed by Reactive Violet K-3R and Eosin Y. In case of the mixed dye solution with low organo-clay load, competitive adsorption took place as the adsorption sites on the surface of the adsorbent were insufficient for all the dye molecules in solution. In the binary system of Direct Fast Scarlet and Eosin Y, the former was preferentially adsorbed due to its greater affinity to EPIDMA/bentonite. Consequently, the equilibrium concentration of Eosin Y was comparatably greater because more adsorption sites were preferentially occupied by Direct Fast Scarlet. As the dosage increased, the equilibrium concentration and the uptake of Direct Fast Scarlet per unit mass of the adsorbent decreased. The competitive adsorption between the dyes was weakened as adsorption of Eosin Y increased. By further increase in adsorbent load, the adsorption sites were sufficient enough for all the dye molecules in the mixed dye solutions, and the adsorption capacity of each dye was not influenced by the presence of the other dyes. The energy of adsorption estimated for the mixed dye systems was less than the mean value of the energy of adsorption of the dyes in single dye systems. The clay-dye interactions in all the cases followed second-order kinetics.

The humic acid immobilized polyacrylamide bentonite composites with amine functionality was shown to be effective in adsorption of the cationic dyes, Malachite green, Methylene Blue and Crystal Violet from their aqueous solutions (Anirudhan et al., 2009) with Langmuir adsorption capacities of 656.6 μmol g^{-1} for Malachite Green, 648.4 $\mu molg^{-1}$ for Methylene Blue and 510.4 $\mu molg^{-1}$ for Crystal Violet at pH 6.0. The authors suggested that, at pH ~ 5.0 to 8.0, most carboxylic acid groups of the immobilized humic acid were deprotonated to form carboxylate groups (RCOO−) resulting in a high possibility of electrostatic attraction of cationic dye molecules to active sites, and causing the observed increase in dye adsorption. All the dyes were adsorbed onto the modified clay composite by following second-order kinetics. However, the rate coefficients showed a steady decrease when the initial dye concentration was increased.

2-aminomethylpyridine immobilized-smectite was used for treatment of water spiked with Sumifix Brilliant Orange 3R (Guerra et al., 2011). The sulfonate and sulfato–ethyl–sulfone groups present in the dye possessed negative charges even in highly acidic solutions that might play a key role in the interaction process. The authors also suggested that, the dye species acted as acidic Lewis centers and were covalently bonded to the clay surface. The molecules anchored onto the oxides containing nitrogen basic atoms supported the dye interaction, mainly of soft cations, due to the presence of the basic reactive centers.

Montmorillonite intercalation with diphosphonium salts [para, meta and ortho bis(triphenyl phosphonium methylene)-benzene-dichloride (p-, m- and o-TPhPMB)] showed improved affinity towards dyes, Telon Red, Telon blue and Telon orange from waste water (Makhoukhi et al., 2010). p-TPhPMBMt showed higher efficiency for the uptake of dyes compared to m-TPhPMB-and o-TPhPMB-modified clay. The amounts of dyes retained decreased in the sequence- orange, red and blue. Thus, the types of diphosphonium ions and dye to be retained played key roles in the total interaction process. The hydrophobic interactions of the dye might involve both diphosphonium ions and the remaining non-covered portion of siloxane surface of the clay. The latter must be sufficiently accessible to dye molecules on p-TPhPMB-clay as compared to the two other organo-clays, due to the relatively small cross section of p-diphosphonium ions and their possible perpendicular arrangement within the interlayer space.

The uptake of Reactive Blue 19 on 1,6-diamino hexane modified bentonite yielded Langmuir capacity of 2.15 x 10^{-4} mol g^{-1} (Gök et al., 2010). The incorporation of 1,6-diaminohexane molecules between the silicate layers of bentonite increased the surface area that might play some role in dye adsorption. The second-order rate coefficient increased steadily from 5.69 x 10^{4} to 6.77 x 10^{4} g $mol^{-1}min^{-1}$ with an increase in the solution temperatures from 293 to 313 K. However, the higher dye concentration decreased the rate coefficient, indicating that adsorption of RB19 onto DAH-bentonite was rate-controlled.

Adsorption of Acid Scarlet from aqueous solution on chitosan modified bentonite and chitosan/hexadecyl trimethyl ammonium modified bentonite (Guo et al., 2012) showed that the latter possessed higher Langmuir adsorption capacity (175.44 mg g^{-1}) compared to chitosan-bentonite (102.04 mg g^{-1}) at 298 K. The introduction of chitosan and chitosan/hexadecyl trimethyl ammonium increased the basal spacing of bentonite from 1.41

nm to 1.96 and 2.74 nm, respectively, which might be responsible for higher uptake of the dye.

The adsorption capacity of chitosan/montmorillonite nanocomposite (molar ratio: 5:1) for Congo Red was reported to be higher than the mean values of those of chitosan and montmorillonite (Wang and Wang, 2007). As the molar ratios of chitosan to montmorillonite increased, the mixtures had a tendency to shrink and thus enhanced the capability of chitosan and montmorillonite agglomerate. This can facilitate the separation of the adsorbents from solution, assisting in the removal process. The Langmuir capacity for Congo Red on N,O-carboxymethyl-chitosan/montmorillonite composite (N,O-carboxymethyl-chitosan : montmorillonite = 5:1) (Wang and Wang, 2008) was reported as 74.14 mg g^{-1} at 303 K. The authors suggested that, at higher molar ratio, the amounts of active hydroxyl (–OH), carboxyl (–COOH) and amine ($–NH_2$) groups increased, resulting in an increase in adsorption capacities of Congo Red. The introduction of N,O-carboxymethyl-chitosan in the montmorillonite improved the flocculation capacity of clay enhancing the adsorption capacity of the composite.

In another work, Monvisade and Siriphannon (2009) used chitosan intercalated montmorillonite for adsorption of three different cationic dyes, Basic Blue 9, Basic Blue 66 and Basic Yellow 1. The Langmuir capacities were highest for Basic Blue 66 followed by Basic Blue 9 and Basic Yellow 1. The adsorption efficiency of chitosan was attributed to the interactions between the functional groups of the dyes and chitosan. The adsorption of the dyes reached ~ 46-49 mg g^{-1}, indicating 92–99 wt. % of dye removal. This increase resulted from the synergistic effect of broadened pores and electrostatic interaction between intercalated chitosan and the dyes. The penetration of the dyes into the interlayer space of modified clays was facilitated by the larger pores of the adsorbent.

The adsorptive removal of Bezactiv Orange V-3R on chitosan/montmorillonite membrane (Nesic et al., 2012) increased with increasing amount of montmorillonite in the membranes. The highest dye removal was observed at 40% montmorillonite that was slightly higher than the capacity of membrane with 50% montmorillonite. It was suggested that increasing the mass ratio of montmorillonite to > 50% in chitosan/ montmorillonite composites could not form a stable composite. The $–CONH_2$, $–NH_2$ and –OH groups from chitosan, and Si-O group from montmorillonite were involved in binding of the azo dye. The kinetic experiments revealed that the interactions were taking place in three stages. The first stage attributed to adsorption on the external surface was correlated to the boundary layer diffusion. The second stage was recognized as the intraparticle diffusion. The third stage referred to the final equilibrium stage.

Cellulose acetate/organo-montmorillonite composite was found to adsorb Acid Scarlet G from aqueous system (Zhou et al., 2012). It was highlighted that the organo-montmorillonite components played a dominant role in the adsorption process. By increasing the content of organo-montmorillonite from 10 wt.% to 60 wt.% in the composites, the adsorption capacities of the dye was enhanced from 3.1 to 85.7 mg g^{-1}, respectively.

The interaction of a diazo dye, Evans blue, with rarasaponin modified bentonite was studied by Chandra et al. (2013). The modification improved the uptake of the dye (q_m for natural bentonite 0.167 mmol g^{-1}; for modified bentonite: 0.527 mmol g^{-1}). The excess acyl groups in the surface of rarasaponin-bentonite became positively charged in the presence of any polar solvent, like water. On the other hand, the presence of SO_3 functional groups developed negative charges in the Evans blue. Thus the adsorption process might have occurred by the electrostatic interaction between the positive charge of acyl groups and negative charge of Evans blue.

Kurniawan et al. (2011) also used rarasaponin modified bentonite for Methylene Blue removal. The Langmuir capacities gradually increased from 231.0 to 256.0 mg g^{-1} as the temperature was raised from 303 to 333 K. The modified-bentonite had unoccupied sites of deacylated carbonyl groups. Being a cationic dye, Methylene blue possessed positive charged surface generated from nitrogen center on its organic framework. The double positive charge of nitrogen center available in Methylene Blue structure might have played an important role during the adsorption process. The imbalance of negative charge of the oxygen atom in unoccupied deacylated carbonyl group could be neutralized by N^+ generated in the cyclohexene structure of the dye. The interactions were close to second-order kinetics.

The adsorption of Safranine dye on alginate encapsulated surfactant-modified pillared bentonite was (Lezehari et al., 2010) explained by Langmuir isotherm model with q_m value 705 μmol g^{-1}. The basic mechanism involved in the adsorption process was ion exchange due to the electrostatic interactions between the cation and the anionic carboxyl groups existing in the polymeric matrix of the polysaccharide.

In another study, chitosan-*g*-poly(acrylic acid)/montmorillonite nanocomposite was used as an adsorbent for Methylene Blue (Wang et al., 2008) yielding a high Langmuir adsorption capacity of 1859.0 mg g^{-1}. The anion functional groups of the adsorbent facilitated absorption and trapping of cationic dyes. Moreover, at the experimental pH (~6.5), –COOH groups present in acrylate were shown to have dissociated to form COO–, increasing the number of fixed ionized groups. This facilitated the creation of electrostatic repulsion forces among the adjacent ionized groups of polymer networks, inducing an expansion of the polymer chains within the nanocomposite structure, which also boosted the absorption for Methylene Blue. The second order kinetic model was favoured by the adsorption process with a rate coefficient of 3.874×10^{-4} mg $g^{-1}min^{-1}$.

Adsorption of Phenols, Substituted Phenols and other Organic Pollutants

The adsorption capacities of some organo-bentonites for phenolic substances, as reported in the literature during the last few years, are listed in Table 4 and the rate coefficients of the adsorption interactions are given in Table 5. The experimental conditions vary considerably from one group of workers to another and the important parameters are shown in the tables of data.

Adsorption data for phenol on montmorillonite modified by intercalation with hexadecyl-trimethyl-ammonium-, polymeric Fe/hexadecyl-trimethyl-ammonium-, polymeric Al/hexadecyl-trimethyl-ammonium- and polymeric Fe/Al/hexadecyl-trimethyl-ammonium- have been shown to follow Freundlich isotherm model (Jiang et al., 2002). The polymeric Al/hexadecyl-trimethyl-ammonium- and hexadecyl-trimethyl-ammonium-modified montmorillonites exhibited improved adsorption of phenol, indicating that these montmorillonites were more hydrophobic in nature than the starting and polymeric Al/Fe-modified montmorillonites. It is noticeable that specific chemical reactions occurred between the phenol and the intercalated adsorption sites resulting in an increase in the adsorption capacity.

Adsorption of phenol on cetyltrimethylammonium bromide-bentonite yielded Langmuir capacity of 333.0 mg g^{-1} at pH 9.0 (Senturk et al. (2009). The organo-bentonite had considerable number of heterogeneous pores where phenol could be trapped and adsorbed. The interaction yielded the second order rate coefficient of 0.116 g mg^{-1} min^{-1}. Adsorption capacities of cetyltrimethylammonium-bentonite towards phenol, aniline, nitrobenzene and naphthalene (Zhu and Zhu., 2008) were controlled by a partition mechanism. The intercalated surfactants were proposed to have formed various partition phases with quite different affinities towards organic solutes as their conformations changed. The adsorption capacities were, however, not reported.

The interactions of tetraethylammonium-bentonite with p-chlorophenol yielded Freundlich capacity of 0.044 mol g^{-1} at 298 K (Akçay, 2004). It was suggested that a donor-acceptor adduct mechanism might be involved in p-chlorophenol adsorption on organo-bentonite. The modified clays acted as Lewis acids, while phenolic compounds in aqueous solution could be viewed as Lewis bases and Brønsted acids.

Adsorption of phenol on montmorillonite modified with hexadecyltrimethyl-ammonium bromide was followed with X-ray diffraction, infrared and near infrared spectroscopy by Liu et al. (2009). The surfactant molecules and phenol molecules entered into the interlayer space of organo-clays, replacing the interlayer cations, as observed from the infrared spectra. It was observed that higher surfactant concentration led to a shift in the IR bands of the organo-clays towards lower wave-numbers and to increased intensity. The near infrared spectra (9000-4000 cm^{-1}) showed a prominent band at 8260 cm^{-1}, arising as a result of combination of CH stretching vibrations of high concentration of hexadecyltrimethylammonium bromide and adsorbed phenol. It is inferred that higher surfactant concentration enhanced phenol adsorption.

Adsorptive accumulation of phenol on bentonite modified with hexadecyl-trimethylammonium bromide (Ocampo-Perez et al., 2011) occurred preferably at a lower temperature with the Langmuir capacity decreasing from 193.0 to 151.0 mg g^{-1} in the temperature range of 288 to 308 K, indicating exothermic interactions. The surface diffusion model was found to fit the experimental data quite well, revealing that the overall rate of adsorption was controlled by surface diffusion. Moreover, the surface diffusion coefficient increased raising the mass of phenol adsorbed at equilibrium and was independent of the particle diameter in the range 0.042-0.0126 cm. The surface diffusion coefficient, for phenol in the organobentonite was in the range of from 4.05 x 10^{-8} to 5.77 x 10^{-8} cm^2/s. The interactions followed second-order kinetics with a rate coefficient of 7.30 x10^{-3} g min^{-1} mg^{-1}.

Adsorption of phenol on hexadecyltrimethylammonium (HDTMA)-bentonite and phenyltrimethylammonium (PTMA)-bentonite (Alkaram et al., 2009) showed that the Langmuir capacities were influenced by solution temperature as well as pH. Higher pH was found to be preferable for adsorption on both the organo-clays. The surfactants covered the bentonite surface with positive charges and increased electrochemical interactions with phenol as it dissociated to $C_6H_5O^-$ at pH > p*K*a (p*K*a = 9.92). The negatively charged species were easily adsorbed on the positive clay surface prevailing at higher pH. The high adsorption capacity might be due to the strong electrostatic interactions between N^+ sites of organo-clays and phenoxide ions. The organo-clays modified with short chain organic cations, like PTMA, possessed limited adsorption capacity due to the finite surface area available; while the clays modified with long chain modifiers, like HDTMA, had no such limitation. Thus, HDTMA-bentonite showed higher Langmuir adsorption capacities in any experimental conditions.

However, the second order rate coefficient for PTMA-bentonite (0.675 g $mg^{-1}min^{-1}$) was higher compared to HDTMA-bentonite (0.128 g $mg^{-1}min^{-1}$).

Hexadecyltrimethylammonium-bentonite-phenol interactions (Mirmohamadsadeghi et al., 2012) yielded type III isotherms which suggested cooperative adsorption such that molecules adsorbed already had a tendency to enhance adsorption of other molecules. Thus, the adsorbate-adsorbate interactions promote adsorption of more molecules leading to formation of several phenol adlayers. The kinetics agreed with a second order rate process with a rate coefficient of 0.365 g $mg^{-1}s^{-1}$.

Adsorption of phenol, p-chlorophenol, and 2,4-dichlorophenol was tested on hexadecyltrimethylammonium-bentonite in aqueous solutions (Rawajfih and Nsour, 2006). The interactions conformed to Freundlich adsorption isotherm in the order of 2,4-dichlorophenol > p-chlorophenol > phenol. It was suggested that adsorption occurred via a dual mode mechanism of partitioning of organic compounds into the organo-clay surface followed by adsorption. The adsorption was influenced by the properties of phenolic compounds. Higher adsorption of 2,4-dichlorophenol might be attributed to its lower water solubility compared to that of the other two phenolic compounds.

Zaghouane-Boudiaf and Boutahala (2011) compared the adsorption efficiency of hexadecyltrimethyl-ammonium-montmorillonite and acid-activated hexadecyltrimethyl-ammonium-montmorillonite for removal of 2,4,5-trichlorophenol from aqueous system. The acid activation did not improve the uptake capacity significantly although the adsorption efficiencies of both the clays were high (q_m for HDTMA-montmorillonite: 368 mg g^{-1}; acid-activated HDTMA-montmorillonite: 303 mg g^{-1}). 2,4,5-trichlorophenol was adsorbed on the organo-montmorillonite by weak van der Waals type of attractive forces to alkyl chains (partition process). In addition, interactions between protonated aluminol groups ($AlOH^{2+}$) in montmorillonite and anionic species of trichlorophenol also took place (ion exchange). The interactions of phenolic compounds with water (hydrogen bonding) were missing or weak as the number of chlorines on the phenol structure increased in water solvent. Thus, it was attracted sufficiently to the hydrophobic surfaces of organo-clays. The interactions of polar organic compounds and organo-clays were governed by adsorption and partition processes, associated with the interfacial behaviour of organic compounds in the organo-clays/water system, and were related to the organic matter and the mineral surface respectively. The experimental data indicated that the adsorption kinetics followed pseudo-second-order rate.

In another study, adsorption of tannin on HDTMA-bentonite (Anirudhan and Ramachandran, 2006) showed a gradual decrease in Langmuir monolayer adsorption capacities from 90.09 μmol g^{-1} (283 K) to 51.81 μmol g^{-1} (313 K). Similar exothermic interactions between HDTMA-bentonite and 2,4,6-trichlorophenol were observed by Anirudhan and Ramachandran (2014) when the Langmuir capacities decreased from 77.51 to 64.10 μmol g^{-1} as the solution temperature was enhanced from 283 to 313 K. The electrostatic attraction between the 2,4,6- trichlorophenol anions and the surfactant cations assisted the adsorbate anions to move into the interlayer space of the clay.

As hexadecyltrimethylammonium ions covered the clay layers completely, the organic medium might be visualized as being formed within the clay layer space and in the interparticle pores. This partition medium boosted the adsorption process. The first order rate coefficients increased gradually and were in the range of 1.24 x 10^{-2} to 1.49 x 10^{-3} min^{-1} (for initial adsorbate concentration of 10 to 75 μmol L^{-1}) and 1.17 x 10^{-2} to 1.64 x 10^{-3} min^{-1} (283 to 313 K).

Table 4. Adsorption capacities of phenol, substituted phenols and other organic pollutants on organo-clays

Organo-clay	Adsorbate	Experimental conditions		Freundlich capacity		Langmuir capacity		Reference
Hexadecyltrimethylammonium-bentonite	Phenol	288 K	Phenol: 50-1200 mg L^{-1}	0.807	$mg^{1-1/n}$ g^{-1} $L^{1/n}$	193.0	mg g^{-1}	Ocampo-Perez et a., 2011
		298 K		0.730		175.8		
		308 K		0.861		151.0		
Hexadecyltrimethylammonium - bentonite	Phenol	298 K	Phenol: 10-100 mg L^{-1}, pH: 6.5	4.499	Lg^{-1}	23.888	mg g^{-1}	Alkaram et al., 2009
		308 K		3.388		19.231		
		318 K		2.720		14.286		
		298 K	Phenol: 10-100 mg L^{-1}, pH: 2.0	3.457		10.173		
			Phenol: 10-100 mg L^{-1}, pH: 6.5	4.499		23.888		
			Phenol: 10-100 mgL^{-1}, pH: 10.0	7.778		25.573		
Hexadecyltrimethylammonium bromide-bentonite	Phenol	303 K	Phenol: 20-1000 mg L^{-1}	0.106	mg g^{-1}			Mirmohamadsadeghi et al., 2012
		313 K		0.155				
		323 K		0.172				
Hexadecyltrimethylammonium-bentonite	Phenol	298 K		0.62	L g^{-1}	105.0	mg g^{-1}	Rawajfih and Nsour, 2006
	p-chloro-phenol			2.36		43.3		
	2,4-dichlo-rophenol			11.88		44.3		
Hexadecyltrimethylammonium-bentonite	2,4,6-trichlo-rophenol	283 K	2,4,6-trichlo-rophenol: 10–400 μmol L^{-1}	18.67	$\mu g^{1-1/n}$ g^{-1} $L^{1/n}$	77.51	μmol g^{-1}	Anirudhan and Rama-chandran., 2014
		293 K		11.68		74.62		
		303 K		9.98		70.42		
		313 K		5.81		64.10		
Hexadecyltrimethylammonium-bentonite	Tannin	283 K		20.60	$\mu g^{1-1/n}$ g^{-1} $L^{1/n}$	90.09	μmol g^{-1}	Anirudhan and Ramachandran, 2006,
		293 K		14.22		77.51		
		303 K		8.71		69.80		
		313 K		5.89		51.81		

Organo-clay	Adsorbate	Experimental conditions		Freundlich capacity		Langmuir capacity		Reference
Hexadecyltrimethylammonium-bentonite	Amoxicillin	293 K		8.64		26.18	mg g^{-1}	Zha et al., 2013
		303 K		8.81		27.85		
		313 K		9.71		30.12		
Hexamethylene bis-pyridinium dibromide-bentonite	Aniline	298 K		0.0336	$mg^{1-1/n}$ g^{-1} $L^{1/n}$			Gu et al., 2014
		308 K		0.0156				
		318 K		0.0109				
		328 K		0.0069				
Hexadecyltrimethylammonium-montmorillonite	Benzoic acid	298 K				26.316	mg g^{-1}	Yildiz et al., 2005
		313 K				39.683		
			pH: 3.0			27.174		
			pH: 6.0			9.488		
			pH: 11.0			6.658		
Hexadecyltrimethylammonium-montmorillonite	Hydroqui-none	298 K				16.639		
		313 K				21.551		
			pH: 3.0			22.523		
			pH: 6.0			17.889		
			pH: 11.0			8.292		
Hexadecyltrimethylammonium-montmorillonite	2,4,5-trichloro-phenol	293±1 K	2,4,5-tri-chlorophenol: 20-30 mgL^{-1}, pH:4.0	15.0	mg g^{-1}	368	mg g^{-1}	Zaghouane-Boudiaf and Boutahala.,2011
Acid-activated hexadecyl trimethylammonium-montmorillonite				13.0		303		
Hexdecyltrimethylammonium-montmorillonite	Phenol		Phenol: 10-100 mgL^{-1}	0.62				Jiang et al., 2002
Fe/hexdecyltrimethylammonium-montmorillonite				0.27				
Al/hexdecyltrimethyl-ammonium-montmorillonite				1.24				

Table 4. (Continued)

Organo-clay	Adsorbate	Experimental conditions		Freundlich capacity		Langmuir capacity		Reference
Fe/Al/hexdecyltrimethylammonium-montmorillonite				0.22				
Cetyltrimethylammonium-bentonite	Phenol		Phenol: 100-1000 mgL^{-1}, pH 9.0	0.099		333.0	mg g^{-1}	Senturk et al., 2009
Ooctadecyltrimethylammonium-montmorillonite	Benzoic acid	298 K				28.486	mg g^{-1}	Yildiz et al., 2005
		313 K				43.290		
			pH: 3.0			30.769		
			pH: 6.0			11.792		
			pH: 11.0			9.671		
Ooctadecyltrimethylammonium-montmorillonite	Hydroqui-none	298 K				12.048		
		313 K				18.519		
			pH: 3.0			20.492		
			pH: 6.0			14.225		
			pH: 11.0			5.376		
Octadecyltrimethylammonium-bentonite	Benzoic acid	298 K	Benzoic acid: 2-12 mmol L^{-1}; pH: 3.5	10.79		20.80	mg g^{-1}	Xin et al., 2011
Octadecyltrimethylammonium/ hydroxyl aluminum-bentonite				11.53		22.03		
Tetraethylammonium-bentonite	P-chloro-phenol	298 K		0.044	mol g^{-1}	-		Akçay, 2004
Amine-modified polyacrylamide–bentonite composite	Humic acid	293 K	Humic acid: 10-400 μmol L^{-1}	14.34		80.12	mg g^{-1}	Anirudhan et al., 2008
		303 K		16.97		103.91		
		313 K		21.39		119.39		
		323 K		27.90		121.64		
Alginate encapsulated surfactant-modified pillared bentonite	Pentachlorophenol	295 K	Adsorbent: 0.08-0.5 g	33.38	μmol$^{1-1/n}$ L$^{1/n}$ g^{-1}	651.0	μmol g^{-1}	Lezehari et al., 2010

Table 5. Second-order rate coefficient (k_2) for adsorption of phenol, substituted phenols and other organic pollutants on organo-clays

Organo-clay	Dye	Experimental conditions			k_2		Reference
Hexadecyltrimethylammonium bromide-bentonite	Phenol		Phenol: 50-1200 mg L^{-1}		7.30 $x10^{-3}$	g mg^{-1} min^{-1}	Ocampo-Perez et a., 2011
Hexadecyltrimethylammonium bromide-bentonite	Phenol		Phenol: 100 mg L^{-1}		0.365	g mg^{-1} s^{-1}	Mirmohamadsadeghi et al., 2012
Hexadecyltrimethylammonium-bentonite	Amoxicillin	303 K			0.0187	g mg^{-1} min^{-1}	Zha et al., 2013
		303 K			0.0189		
		313 K			0.0225		
Hexadecyltrimethylammonium-montmorillonite	2,4,5-trichloro-phenol	293 K	Clay 20 mg	pH: 2.0	1.84 $x10^3$	g mg^{-1} min^{-1}	Zaghouane-Boudiaf and Boutahala., 2011
				pH: 4.0	2.21 $x10^3$		
				pH: 6.0	0.91 $x10^3$		
				pH: 8.0	8.50 $x10^3$		
			Clay 30 mg	pH: 4.0	2.35 $x10^3$		
			Clay 40 mg		2.03 $x10^3$		
		303 K	Clay 20 mg	pH: 4.0	1.64 $x10^3$		
		313 K			1.00 $x10^3$		
Hexadecyltrimethylammonium - bentonite	Phenol	298 K	Phenol 10-100 mg L^{-1}, pH: 6.5		0128	g mg^{-1} min^{-1}	Alkaram et al., 2009
Phenyltrimethylammonium-bentonite					0.675		
Acid-activated hexadecyl trimethylammonium-montmorillonite	2,4,5-trichloro-phenol	293 K	Clay 20 mg	pH: 2.0	23.00 $x10^3$	g mg^{-1} min^{-1}	Zaghouane-Boudiaf and Boutahala., 2011
				pH: 4.0	6.82 $x10^3$		
				pH: 6.0	20.00 $x10^3$		
				pH: 8.0	21.11 $x10^3$		
			Clay 30 mg	pH: 4.0	4.52 $x10^3$		
			Clay 40 mg		7.11 $x10^3$		
		303 K	Clay 20 mg	pH: 4.0	2.73 $x10^3$		
		313 K			2.00 $x10^3$		

Table 5. (Continued)

Organo-clay	Dye	Experimental conditions		k_2		Reference
Cetyltrimethylammonium bromide-bentonite	Phenol		Phenol: 100-1000 mg L^{-1}, pH 9.0	0.116	g mg^{-1} min^{-1}	Senturk et al., 2009
Amine-modified polyacrylamide – bentonite composite	Humic acid	293 K	Humic acid: 100 µmolL^{-1}, pH:6.0	1.0 x10^{-3}	µmol g^{-1} min^{-1}	Anirudhan et al., 2008
		303 K		1.2 x10^{-3}		
		313 K		1.5 x10^{-3}		
		323 K		1.8 x10^{-3}		
		303 K	Humic acid 25 µmol L^{-1}	4.2 x10^{-3}		
			Humic acid 50 µmol L^{-1}	2.1 x10^{-3}		
			Humic acid 75 µmol L^{-1}	1.5 x10^{-3}		
			Humic acid 100 µmol L^{-1}	1.2 x10^{-3}		
Alginate encapsulated surfactant-modified pillared bentonite	Pentachlorophenol	295 K		0.023	µmol h	Lezehari et al., 2010

The adsorption of amoxicillin on hexadecyltrimethylammonium-bentonite (Zha et al., 2013) preferred higher temperature as Langmuir capacity increased from 26.18 to 30.12 mg g^{-1}. It was suggested that, the mechanism for the adsorption of amoxicillin the mechanism for the adsorption of amoxicillin depended not only on the partition mechanism but also on the hydrocarbon chain length, degree of exchangeable cation replacement by surfactant molecules and the number of hydrocarbon chains within the surfactant molecule. The second order rate coefficient was in the range of 0.0187 to 0.0225 g mg^{-1} min^{-1}.

Gu et al. (2014) used hexamethylene bis-pyridinium dibromide modified bentonite for removal of aniline. The interactions were exothermic in nature as the Freundlich capacity decreased from 0.0336 to 0.0069 $mg^{1-1/n}$ g^{-1} $L^{1/n}$ on increasing the solution temperature from 298 to 328 K. At experimental pH (= 6.6), aniline existed in the form of molecules, and the surface charge of organo-clay could not influence its adsorption. The π-π polar interactions between the aromatic rings were supposed to be responsible for the enhanced adsorption of aniline on hexamethylene bis-pyridinium dibromide-bentonite.

Xu and Zhu (2007) compared phenol adsorption on hexamethonium-bentonite with tetramethyl ammonium-bentonite. Being a diammonium cation with a short-alkyl chain, hexamethonium bromide was shown to have intercalated in the clay interlamellar space and to have lied flat on the siloxane surface forming hydrocarbon surface and leaving free siloxane mineral surfaces. Therefore, the dominant mechanism in the uptake of phenol on hexamethonium-bentonite was adsorption similar to that of tetramethyl ammonium-bentonite.

The comparison of phenol adsorption on phenyltrimethylammonium-and benzyltrimethylammonium-bentonite (Majdan et al., 2009) indicated a clear difference for phenol adsorption isotherms. The adsorption isotherm for phenyltrimethylammonium-bentonite was of straight-line character resulted from simple partitioning of phenol between the aqueous and organic phases sorbed on the bentonite surface. For benzyltrimethylammonium-bentonite the isotherm had a convex shape, characteristic of physicochemical adsorption. The adsorption of phenol took place via interactions between the phenyl rings of phenol molecules and organo-clay complex, however, these interactions were stronger and affected the strength of the hydrogen bonding between phenol and water in benzyltrimethylammonium-bentonite. On the other hand, the dimerization of phenyltrimethylammonium- cations played a very important role in weakening of the phenol uptake on phenyltrimethylammonium-bentonite when compared with that of benzyltrimethylammonium-bentonite.

Xu and Zhu (2009) used octadecyltrimethylammonium- and dioctadecyldimethyl ammonium-bentonites for removal of naphthalene and nitrobenzene, though the adsorption capacities were not mentioned. At high loading region, octadecyltrimethylammonium-molecules appeared in the clay interlayer via hydrophobic interactions between alkyl chains. The strong hydration of surfactant ammonium and the counter ions (Cl^-) in aqueous system interfered with the hydrophobic interactions of the octadecyltrimethylammonium- clusters and destroyed the close packing. It was observed that the adsorption capacities of dioctadecyldimethylammonium-bentonite towards naphthalene and nitrobenzene were higher than those of octadecyltrimethylammonium-bentonite.

The potential use of hexadecyltrimethyleammonium- (HDTMA-) and octadecyl trimethyle ammonium- (ODTMA) modified montmorillonite for removal of benzoic acid and hydroquinone (Yildiz et al., 2005) showed that the adsorption capacity of ODTMA-montmorillonite for benzoic acid was higher compared to that of HDTMA- montmorillonite at various pHs and temperatures. The reverse trend was observed for hydroquinone adsorption. The higher aqueous solubility of hydroquinone (7.10 g/100 cm^3) compared to benzoic acid (0.32 g/100 cm^3) might play some role in lowering its uptake by the organo-clays. The uptake of both benzoic acid and hydroquinone decreased with increasing pH. A lower adsorption at higher pH may be due to the abundance of OH^- ions resulting the ionic repulsion between the negatively charged surface and the ionic organic compounds. On the other hand, the increased adsorption at low pH suggested that the adsorption of the less ionic species of benzoic acid and hydroquinone was higher than the more ionic species.

Benzoic acid was also removed by using octadecyl trimethyl ammonium- and octadecyl trimethyl ammonium/hydroxyl aluminum-bentonite (Xin et al. (2011). The Langmuir capacities of octadecyl trimethyl ammonium/hydroxyl aluminum-bentonite was higher (22.03 mg g^{-1}) compared to octadecyl trimethyl ammonium- bentonite (20.80 mg g^{-1}). Both the organo-clays possessed high removal efficiency for benzoic acid at pH ~3.0 to 4.0, while the uptake of benzoic acid decreased with increasing pH. At higher pH, the abundance of OH^- ions competing with the organic anions for the adsorption sites might play some role in lowering the uptake. Moreover, the electrostatic repulsion between the negatively charged clay surface and the ionic organic compounds retarded the removal process.

In the endothermic interactions of amine-modified polyacrylamide–bentonite composite with humic acid (Anirudhan et al., 2008), it was observed that the Langmuir adsorption capacities increased from 80.12 to 121.64 mg g^{-1} when the solution temperature was raised from 293 to 323 K. At pH 6.0, the surface of the organo-clay was electrically positive and the –COOH groups of humic acid dissociated into HA-COO^-, adsorption was caused by strong electrostatic interactions. The adsorptive interactions were viewed as the complexation of the carboxyl groups in humic acid with the protonated NH_2 groups of the organo-clay. An increase in initial humic acid concentration (25 to 100 μmol L^{-1}) decreased the second-order rate coefficient from 4.2×10^{-3} to 1.2×10^{-3} μmol g^{-1} min^{-1} indicating that the initial adsorbate concentration could influence the equilibrium time. Due to the endothermic nature of the interactions, higher temperature boosted the rate coefficient from 1.0×10^{-3} μmol g^{-1} min^{-1} (293 K) to 1.8×10^{-3} μmol g^{-1} min^{-1} (323 K). The authors have also reported a significant role for the film diffusion mechanism in the adsorption process.

Lezehari et al. (2010) used alginate encapsulated surfactant-modified pillared bentonite as the adsorbent for pentachlorophenol in water. The Langmuir capacity was 651.0 μmol g^{-1} at 295 K. The formation of pentachlorophenolate at pH 5.3 might have been responsible for uptake of the adsorbate. The ion exchange due to the electrostatic interactions between the cation and the anionic carboxyl groups existing in the polymeric matrix of the polysaccharide was the main mechanism of the adsorptive interactions. The process was governed by second-order kinetics with a rate coefficient of 0.023 μmol. h^{-1} at 298 K.

Conclusion and Future Perspectives

In this review, an attempt has been made to focus on the recent developments (for the last 10 years) related to the detoxification of water and wastewater from toxic dyes, phenols and substituted phenols, and some other organics by organically modified montmorillonite. A good volume of work has resulted on the use of organo-montmorillonite with a steadily increasing upward swing during the last few years, even as many areas of knowledge-gap are yet to be addressed. A few of the important issues which need to be addressed are summarized below:

(i) The cost of chemical modification of the clays with various organic surfactants and others has to be considered and the same will have to be justified on the basis of higher adsorption efficiency.
(ii) The efficiency of organo-montmorillonite in multi-component systems needs to be explored.
(iii) In order to have practical significance, some work needs to be undertaken with actual contaminated water under continuous flow system.
(iv) Desorption and regeneration studies have not received much attention with the organo-clays and some efforts should be given on recovery/reuse and disposal of the pollutant-loaded organo-montmorillonite.

Note: Different group of authors used different units for Freundlich and Langmuir adsorption capacities. The units are given (both in text and table) as these were mentioned in the published works that have been reviewed.

REFERENCES

Abou Taleb, M. F., Hegazy, D. E. & Ismail, S. A. (2012). Radiation synthesis, characterization and dye adsorption of alginate-organophilic montmorillonite nanocomposite, *Carbohydrate Polymers*, *87*, 2263– 2269.

Ahmaruzzaman, M. (2008). Adsorption of phenolic compounds on low-cost adsorbents: A review, *Advances in Colloid and Interface Science*, *143*, 48–67.

Akcay, G., Akcay, M. & Yurdakoc, K. (2006). The characterization of prepared organomontmorillonite (DEDMAM) and sorption of phenoxyalkanoic acid herbicides from aqueous solution, *Journal of Colloid and Interface Science*, *296*, 428 – 433.

Akçay, M. (2004). Characterization and determination of the thermodynamic and kinetic properties of *p*-CP adsorption onto organophilic bentonite from aqueous solution, *Journal of Colloid and Interface Science*, *280*, 299–304.

Alkaram, U. F., Mukhlis, A. A. & Al-Dujaili, A. H. (2009). The removal of phenol from aqueous solutions by adsorption using surfactant-modified bentonite and kaolinite, *Journal of Hazardous Materials*, 169, 324–332.

Allen, S. J. & Koumanova, B. (2005). Decolourisation of water/wastewater using adsorption (review), *Journal of the University of Chemical Technology and Metallurgy*, *40*, 3, 175-192.

Anirudhan, T. S. & Ramachandran, M. (2006). Adsorptive removal of tannin from aqueous solutions by cationic surfactant-modified bentonite clay, *Journal of Colloid and Interface Science*, *299*, 116–124.

Anirudhan, T. S. & Ramachandran, M. (2014). Removal of 2,4,6-trichlorophenol from water and petroleum refinery industry effluents by surfactant-modified bentonite, *Journal of Water Process Engineering*, *1*, 46–53.

Anirudhan, T. S., Suchithra, P. S. & Radhakrishnan, P. G. (2009). Synthesis and characterization of humic acid immobilized-polymer/bentonite composites and their ability to adsorb basic dyes from aqueous solutions, *Applied Clay Science*, *43*, 336–342.

Anirudhan, T. S., Suchithra, P. S. & Rijith, S. (2008). Amine-modified polyacrylamide–bentonite composite for the adsorption of humic acid in aqueous solutions, *Colloids and Surfaces A: Physicochem. Eng. Aspects*, *326*, 147–156.

Arellano-Cárdenas, S., López-Cortez, S., Cornejo-Mazón, M. & Mares-Gutiérre, J. C. (2013). Study of malachite green adsorption by organically modified clay using a batch method, *Applied Surface Science*, *280*, 74–78.

ATSDR, (1999). Toxicological profile for chlorophenol, Agency for Toxic Substances and Disease Registry (http://www.atsdr.cdc.gov), U.S. Department of Health and Human Services, Atlanta, Georgia.

ATSDR, (2008). Toxicological profile for phenol, Agency for Toxic Substances and Disease Registry (website: http://www.atsdr.cdc.gov).

Balomenou, G., Stathi, P., Enotiadis, A., Gournis, D. & Deligiannakis, Y. (2008). Physicochemical study of amino-functionalized organosilicon cubes intercalated in montmorillonite clay: H-binding and metal uptake, *Journal of Colloid and Interface Science*, *325*, 74–83.

Baskaralingam, P., Pulikesi, M., Elango, D., Ramamurthi, V. & Sivanesan, S. (2006). Adsorption of acid dye onto organobentonite, *Journal of Hazardous Materials*, B128, 138–144.

Bhatnagar, A. & Sillanpää, M. (2010). Utilization of agro-industrial and municipal waste materials as potential adsorbents for water treatment - A review, *Chemical Engineering Journal*, *157*, 277–296.

Bhatnagar, A., Vilar, V. J. P., Botelho, C. M. S. & Boaventura, R. A. R. (2010). Coconut-based biosorbents for water treatment - A review of the recent literature, *Advances in Colloid and Interface Science*, *160*, 1–15.

Bhatt, A. S., Sakaria, P. L., Vasudevan, M., Pawar, R. R., Sudheesh, N., Bajaj, H. C. & Mody, H. M. (2012). Adsorption of an anionic dye from aqueous medium by organoclays: equilibrium modeling, kinetic and thermodynamic exploration, *RSC Advances*, *2*, 8663–8671.

Bhattacharyya, K. G. & Sen Gupta, S. (2008). Adsorption of a few heavy metals on natural and modified kaolinite and montmorillonite: A review, *Advances in Colloid and Interface Science*, *140*, 114–131.

Bhattacharyya, K. G. & Sen Gupta, S. (2009). Calcined tetrabutylammonium kaolinite and montmorillonite and adsorption of Fe(II), Co(II) and Ni(II) from solution, *Applied Clay Science*, *46*, 216–221.

Bianchi, A. E., Fernández, M., Pantanetti, M., Viña, R., Torriani, I., Torres Sánchez, R. M., Punte, G. (2013). $ODTMA^+$ and $HDTMA^+$ organo-montmorillonites characterization: New insight by WAXS, SAXS and surface charge, *Applied Clay Science*, 83-84, 280-285.

Bradley, W. F. & Grim, R. E. (1948). Colloidal properties of layer silicates, *J. Phys. Colloid Chem.*, *52*, 1404-1413.

Brigati, M. F., Galan, E. & Theng, B. K. G. (2006). Structure and mineralogy of clay minerals, In: *Handbook of Clay Science* (Eds. F. Bergaya, B. K. G. Theng, G. Lagaly), chapter 2, Elsevier, 19-86.

Carvalho, W. A., Vignado, C. & Fontana, J. (2008). Ni(II) removal from aqueous effluents by silylated clays, *Journal of Hazardous Materials*, *153*, 1240-1247.

Chandra, I. K., Ju, Y. H. Ayucitra, A. & Ismadji, S. (2013). Evans blue removal from wastewater by rarasaponin-bentonite, *Int. J. Environ. Sci. Technol.*, *10*, 359-370.

Crains–Smith, A. G. & Hartman, H. (Eds.), (1986). *Clay Minerals and the Origin of Life*, Cambridge Univ. Press, Cambridge.

Crini, G. (2008). Kinetic and equilibrium studies on the removal of cationic dyes from aqueous solution by adsorption onto a cyclodextrin polymer, *Dyes and Pigments*, *77*, 415–426.

Deer, W. A., Howie, R. A., Zussman, J. (1985). *An Introduction to the Rock-Forming Minerals,* ELBS Longman, Essex, England, 250-254.

Elemen, S., Kumbasar, E. P. A. & Yapar, S. (2012). Modeling the adsorption of textile dye on organoclay using an artificial neural network, *Dyes and Pigments*, *95*, 102-111.

Erdem, B., Özcan, A., Gök, Ö. & Özcan, A. S. (2009). Immobilization of 2, 2'-dipyridyl onto bentonite and its adsorption behaviour of copper(II) ions, *Journal of Hazardous Materials*, *163*, 418-426.

Fewson, C. A. (1998). Biodegradation of xenobiotic and other persistent compounds:the causes of recalcitrance. *Trends Biotechnol.*, *6*, 148-153.

Freundlich, H. M. F. (1906). Over the adsorption in solution, *J. Phys. Chem.*, *57*, 385-470.

Gök, Ö., Özcan, A. S. & Özcan, A. (2010). Adsorption behavior of a textile dye of Reactive Blue 19 from aqueous solutions onto modified bentonite, *Applied Surface Science*, *256*, 5439–5443.

Gök, Ö., Özcan, A., Erdem, B. & Özcan, A. S. (2008). Prediction of the kinetics, equilibrium and thermodynamic parameters of adsorption of copper(II) ions onto 8-hydroxy quinoline immobilized bentonite, *Colloids and Surfaces A: Physicochem. Eng. Aspects*, *317*, 174-185.

Grim, R.E. (1962). *Applied Clay Mineralogy, International Series in the Earth Sciences*, McGraw Hill, New York.

Gu, Z., Gao, M., Luo, Z., Lu, L., Ye, Y. & Liu, Y. (2014). Bis-pyridinium dibromides modified organo-bentonite for the removal of aniline from wastewater: A positive role of π-π polar interaction, *Applied Surface Science*, *290*, 107-115.

Guerra, D. L., Oliveira, H. C. P., da Costa, P. C. C., Viana, R. R. & Airoldi, C. (2010). Adsorption of chromium(VI) ions on Brazilian smectite: Effect of contact time, pH, concentration, and calorimetric investigation, *Catena*, *82*, 35–44.

Guerra, D. L., Silva, W. L. L., Oliveira, H. C. P., Viana, R. R. & Airoldi, C. (2011). Organofunctionalized Amazon smectite for dye removal from aqueous medium-Kinetic and thermodynamic adsorption investigations, *Journal of Hazardous Materials*, *186*, 675–682.

Guggenheim, S. & Martin, R. T. (1995). Definition of clay and clay mineral: joint report of the AIPEA nomenclature and CMS nomenclature committees, *Clays and Clay Minerals*, *43*(2), 255-256.

Guo, J., Chen, S., Liu, L., Li, B., Yang, P., Zhang, L. & Feng, Y. (2012). Adsorption of dye from wastewater using chitosan–CTAB modified bentonites, *Journal of Colloid and Interface Science*, *382*, 61–66.

Gupta, V. K., Carrott, P. J. M., Ribeiro Carrott, M. M. L. & Suhas, (2009). Low-cost adsorbents: growing approach to wastewater treatment-a review, *Critical Reviews in Environmental Science and Technology*, *39*, 783–842.

Hao, O. J., Kim, H. & Chian, P. C. (2000). Decolorization of wastewater, *Critical Reviews in Environmental Science and Technology*, *30*, 449–505.

Hendricks, S. B. (1942). Lattice structure of clay minerals and some properties of clays, *J. Geol.*, *50*, 276-290.

Ho, Y. S. (2004). Citation review of Lagergren kinetic rate equation on adsorption reactions, *Scientometrics*, *59*, 171 – 177.

Ho, Y. S. & McKay, G. (1999a). Batch lead (II) removal from aqueous solution by peat: Equilibrium and kinetics, *Trans I Chem E.*, *77*B, 165-173.

Ho, Y. S. & McKay, G. (1999b). The sorption of lead(II) ions on peat, *Water Research*, *33*, 578 –584.

Huang, M. & Zhu, R. (2011). Sorption of 2,4-dichlorophenol onto organobentonites: influence of organic cation structure and bentonite layer charge, *Adsorption Science & Technology*, *29*, 29-38.

Jamirah, A. I. & Dental, S. K. (1993). Mechanisms of reactions between montmorillonite and quaternary amines salts, Joint CSCE-ASCE National conference on Environmental Engineering, July, 12-14, Montreal, Canada, 1065-1072.

Jiang, J. Q., Cooper, C. & Ouki, S. (2002). Comparison of modified montmorillonite adsorbents Part I: preparation, characterization and phenol adsorption, *Chemosphere*, *47*, 711–716.

Jing, P., Hou, M., Zhao, P., Tang, X. & Wan, H. (2013). Adsorption of 2-mercaptobenzothiazole from aqueous solution by organo-bentonite, *Journal of Environmental Sciences*, *25*, 1139–1144.

Jović-Jovičić, N. P., Milutinović-Nikolić A. D., Žunić, M. J., Mojović, Z. D., Banković, P. T., Gržetić, I. A. & Jovanović, D. M. (2013). Synergic adsorption of Pb^{2+} and reactive dye — RB5 on two series of organomodified bentonites, *Journal of Contaminant Hydrology*, *150*, 1–11.

Kang, Q., Zhou, W., Li, Q., Gao, B., Fan, J. & Shen, D. (2009). Adsorption of anionic dyes on poly(epicholorohydrin dimethylamine) modified bentonite in single and mixed dye solutions, *Applied Clay Science*, *45*, 280–287.

Kaushik, P. & Malik, A. (2009). Fungal dye decolorization: Recent advances and future potential, *Environ. Int. 35*, 127-141.

Koswojo, R., Utomo, R. P., Ju, Y. H., Ayucitra, A., Soetaredjo, F. E., Sunarso, J. & Ismadji, S. (2010). Acid Green 25 removal from wastewater by organo-bentonite from Pacitan, *Applied Clay Science*, 48, 81–86.

Koyuncu, H., Yildiz, N., Salgın, U., Köroğlu, F. & Çalimli, A. (2011). Adsorption of *o-*, *m-* and *p*-nitrophenols onto organically modified bentonites, *Journal of Hazardous Materials*, 185, 1332-1339.

Kurniawan, A., Sutiono, H., Ju, Y. H., Soetaredjo, F. E., Ayucitra, A., Yudha, A. & Ismadji, S. (2011). Utilization of rarasaponin natural surfactant for organo-bentonite preparation: Application for methylene blue removal from aqueous effluent, *Microporous and Mesoporous Materials*, *142*, 184-193.

Lagaly, G. (2006). Colloid clay science, In: *Handbook of Clay Science* (Eds. F. Bergaya, B. K. G. Theng, G. Lagaly), chapter 5, Elsevier, 141-245.

Lagaly, G. & Ogawa, M. (2006). Dékány, Clay mineral organic interactions, In: *Handbook of Clay Science* (Eds. F. Bergaya, B. K. G. Theng, G. Lagaly), chapter 7.3, Elsevier, 309-371.

Lagaly, G., Witter, R. & Sander, H. (1983). Water on hydrophobic surfaces, In: *Adsorption from Solution* (Eds. R. H. Ottewill, C. H. Rochester), Academic Press, London, 65-77.

Lagergren, S. (1898). Zur theorie der sogenannten adsorption gelöster stoffe, *Kungliga Svenska Vetenskapsakademiens,* Handlingar, Band *24*, No. 4, 1-39.

Langmuir, I. (1918). The adsorption of gases on plane surfaces of glass, mica, and platinum, *Journal of the American Chemical Society*, *40*, 1361-1403.

Lawrence, M. A. M., Kukkadapu, R. K. & Boyd S. A. (1998). Adsorption of phenol and chlorinated phenols from aqueous solution by tetramethylammonium- and tetramethylphosphonium-exchanged montmorillonite, *Applied Clay Science*, *13*, 13-20.

Lee, S. M. & Tiwari, D. (2012). Organo and inorgano-organo-modified clays in the remediation of aqueous solutions: An overview, *Applied Clay Science*, 59-60, 84–102.

Lezehari, M., Basly, J. P., Baudu, M. & Bouras, O. (2010). Alginate encapsulated pillared clays: removal of a neutral/anionic biocide (pentachlorophenol) and a cationic dye (safranine) from aqueous solutions, *Colloids and Surfaces A: Physicochem. Eng. Aspects*, *366*, 88–94.

Li, Q., Su, Y., Yue, Q. Y. & Gao, B. Y. (2011). Adsorption of acid dyes onto bentonite modified with polycations: Kinetics study and process design to minimize the contact time, *Applied Clay Science*, *53*, 760-765.

Li, Q., Yue, Q. Y., Su, Y., Gao, B. Y. & Sun, H. J. (2010). Equilibrium, thermodynamics and process design to minimize adsorbent amount for the adsorption of acid dyes onto cationic polymer-loaded bentonite, *Chemical Engineering Journal*, *158*, 489-497.

Li, Q., Yue, Q. Y., Sun, H. J., Su, Y. & Gao, B. Y. (2010). A comparative study on the properties, mechanisms and process designs for the adsorption of non-ionic or anionic dyes onto cationic-polymer/bentonite, *Journal of Environmental Management*, *91*, 1601-1611.

Lin, S. H. & Juang, R. S. (2009). Adsorption of phenol and its derivatives from water using synthetic resins and low-cost natural adsorbents: A review, *Journal of Environmental Management, 90*, 1336-1349.

Liu, B., Wang, X., Yang, B. & Sun, R. (2011). Rapid modification of montmorillonite with novel cationic Gemini surfactants and its adsorption for methyl orange, *Materials Chemistry and Physics*, *130*, 1220–1226.

Liu, N., Wang, M. X., Liu, M. M., Liu, F., Weng, L., Koopal, L. K. & Tan, W. F. (2012). Sorption of tetracycline on organo-montmorillonites, *Journal of Hazardous Materials*, 225–226, *30*, 28–35.

Liu, P. & Zhang, L. (2007). Adsorption of dyes from aqueous solutions or suspensions with clay nano-adsorbents, *Separation and Purification Technology*, *58*, 32–39.

Liu, R., Frost, R. L. & Martens, W. N. (2009). Near infrared and mid infrared investigations of adsorbed phenol on HDTMAB organoclays, *Materials Chemistry and Physics*, *113*, 707–713.

Ma, J., Cui, B., Dai, J. & Li, D. (2011). Mechanism of adsorption of anionic dye from aqueous solutions onto organobentonite, *Journal of Hazardous Materials*, *186*, 1758–1765.

Maegdefrau, E. & Hofmann, U. (1937). Die kristalstructure des montmorillonite, *Z. Krist.*, *98*, 299-323.

Majdan, M., Bujacka, M., Sabah, E., G1adysz-P1aska, A., Pikus, S., Sternik, D., Komosa, Z. & Padewski, A. (2009). Unexpected difference in phenol sorption on PTMA- and BTMA-bentonite, *Journal of Environmental Management*, *91*, 195–205.

Makhoukhi, B., Didi, M. A., Moulessehoul, H., Azzouz, A. & Villemin, D. (2010). Diphosphonium ion-exchanged montmorillonite for Telon dye removal from aqueous media, *Applied Clay Science*, *50*, 354–361.

Mattson, J. S. & Mark, H. B. (1971). *Activated Carbon: Surface Chemistry and Adsorption from Solution*, Marcel Dekker, New York.

Mering, J. (1946). On the hydration of montmorillonite, *Trans. Faraday Soc.*, *42B*, 205-219.

Mering, J. & Glaeser, R. (1953). Cations echangeables dans montmorillonite, *Bull. Group Franc., Argiles*, *5*, 61-72.

Meshko, V., Markovska, L., Mincheva, M. & Rodrigues, A. E. (2001). Adsorption of basic dyes on granular activated carbon and natural zeolites, *Water Research*, 5, 3357– 3366.

Mirmohamadsadeghi, S., Kaghazchi, T., Soleimani, M. & Asasian, N. (2012). An efficient method for clay modification and its application for phenol removal from wastewater, *Applied Clay Science*, 59–60, 8–12.

Monvisade, P. & Siriphannon, P. (2009). Chitosan intercalated montmorillonite: Preparation, characterization and cationic dye adsorption, *Applied Clay Science*, *42*, 427–431.

Mortland, M. M., Shaobai, S. & Boyd, S. A. (1986). Clay-organic complexes as adsorbents for phenol and chlorophenols, *Clays and Clay Minerals*, *34*, 581– 585.

Nesic, A. R., Velickovic, S. J. & Antonovic, D. G. (2012). Characterization of chitosan/montmorillonite membranes as adsorbents for Bezactiv Orange V-3R dye, *Journal of Hazardous Materials*, 209-210, 256-263.

Özcan, A. S., Erdem, B. & Özcan, A. (2004). Adsorption of Acid Blue 193 from aqueous solutions onto Na-bentonite and DTMA-bentonite, *Journal of Colloid and Interface Science*, *280*, 44–54.

Rangabhashiyam, S., Anu, N. & Selvaraju, N. (2013). Sequestration of dye from textile industry wastewater using agricultural waste products as adsorbents. *J. Environ. Chem. Eng.*, *1*, 629–641.

Rawajfih, Z. & Nsour, N. (2006). Characteristics of phenol and chlorinated phenols sorption onto surfactant-modified bentonite, *Journal of Colloid and Interface Science*, *298*, 39–49.

Salgın, U., Yildiz, N., Köroğlu, F. & Çalimli, A. (2007). A parametric study of phenolic compounds desorption performance from organobentonite by supercritical fluids, *Journal of Supercritical Fluids*, *39*, 296–303.

Sen Gupta, S. & Bhattacharyya, K. G. (2012). Adsorption of heavy metals on kaolinite and montmorillonite: a review, *Phys. Chem. Chem. Phys.*, *14*, 6698–6723.

Sen Gupta, S. & Bhattacharyya, K. G. (2014). Adsorption of metal ions by clays and inorganic solids, *RSC Advances*, *4*, 28537-28586.

Senturk, H. B., Ozdes, D., Gundogdu, A., Duran, C. & Soylak, M. (2009). Removal of phenol from aqueous solutions by adsorption onto organomodified Tirebolu bentonite: Equilibrium, kinetic and thermodynamic study, *Journal of Hazardous Materials*, *172*, 353–362.

Shen, D., Fan, J., Zhou, W., Gao, B., Yue, Q. & Kang, Q. (2009). Adsorption kinetics and isotherm of anionic dyes onto organo-bentonite from single and multisolute systems, *Journal of Hazardous Materials*, *172*, 99–107.

Sheng, G. & Boyd, S. A. (2000). Polarity effect on dichlorobenzene sorption by hexadecyltrimethyl-ammonium exchanged clays, *Clays and Clay Minerals*, *48*, 43-50.

Silva, M. M. F., Oliveira, M. M., Avelino, M. C., Fonseca, M. G., Almeida, R. K. S. & Silva Filho, E. C. (2012). Adsorption of an industrial anionic dye by modified-KSF-montmorillonite: Evaluation of the kinetic, thermodynamic and equilibrium data, *Chemical Engineering Journal*, *203*, 259–268.

Śölener, M., Tunali, S., Özcan, A. S., Özcan, A. & Gedikbey, T. (2008). Adsorption characteristics of lead(II) ions onto the clay/poly(methoxyethyl)acrylamide (PMEA) composite from aqueous solutions, *Desalination*, *223*, 308–322.

Srinivasan, A. & Viraraghavan, T. (2010). Decolorization of dye wastewaters by biosorbents: A review, *Journal of Environmental Management*, *91*, 1915-1929.

Stevens, J. J. & Anderson, S. J. (1996). An FTIR study of water sorption on TMA- and TMPA-montmorillonites, *Clays and Clay Minerals*, *44*, 142-150.

Sun, X., Huang, W., Ma, Z., Lu, Y. & Shen, X. (2013). A novel approach for removing 2-naphthol from wastewater using immobilized organo-bentonite, *Journal of Hazardous Materials*, 252–253, 192–197.

Tanabe, K. (1981). Solid acid and base catalysis; in: *Catalysis – Science and Technology* (Eds. Anderson, J.R., Boudart, M.), Springer – Verlag, New York, 231.

Thomas, W. J. & Crittenden, B. (1998). Adsorbents, in: *Adsorption Technology and Design*, chapter 2, Butterworth–Heinemann, Oxford, 8-30.

Tong, D. S., Zhou, C. H. (C.), Lu, Y., Yu, H., Zhang, G. F.& Yu, W. H. (2010). Adsorption of Acid Red G dye on octadecyl trimethylammonium montmorillonite, *Applied Clay Science*, *50*, 427–431.

Tran H. H., Roddick, F. A. & O'Donnell, J. A. (1999). Comparison of chromatography and desiccant silica gels for the adsorption of metal ions – I. Adsorption and kinetics, *Water Research*, *33*, 2992-3000.

van Olphen, H. (1977). *An Introduction to Clay Colloid Chemistry,* Wiley Interscience, 425-428.

Wang, S. & Peng, Y. (2010). Natural zeolites as effective adsorbents in water and wastewater treatment, *Chemical Engineering Journal*, *156*, 11-24.

Wang, L. & Wang, A. (2007). Adsorption characteristics of Congo Red onto the chitosan/ montmorillonite nanocomposite, *Journal of Hazardous Materials*, *147*, 979–985.

Wang, L. & Wang, A. (2008a). Adsorption properties of Congo Red from aqueous solution onto surfactant-modified montmorillonite, *Journal of Hazardous Materials*, *160*, 173–180.

Wang, L. & Wang, A. (2008b). Adsorption behaviors of Congo Red on the N,O-carboxymethyl-chitosan/montmorillonite nanocomposite, *Chemical Engineering Journal*, *143*, 43–50

Wang, L., Zhang, J. & Wang, A. (2008). Removal of methylene blue from aqueous solution using chitosan-*g*-poly (acrylic acid)/montmorillonite superadsorbent nanocomposite, *Colloids and Surfaces A: Physicochem. Eng. Aspects*, *322*, 47–53.

Weber, W. J. Jr., McGinley, P. M. & Katz, L. E. (1991). Sorption phenomena in subsurface systems: concepts, models and effects on contaminant fate and transport, *Water Research*, *25*, 499-528.

WHO, (2003). Chlorophenols in drinking-water, Background document for preparation of WHO Guidelines for drinking-water quality, Geneva, World Health Organization (WHO/SDE/WSH/03.04/47).

Wibulswas, R. (2004). Batch and fixed bed sorption of methylene blue on precursor and QACs modified montmorillonite, *Separation and Purification Technology*, *39*, 3–12.

Xi, Y., Mallavarapu, M. & Naidu, R. (2009). Preparation, chracterization of surfactants modified clay minerals and nitrate adsorption, *Applied Clay Science, 48*, 92–96.

Xin, X., Si, W., Yao, Z., Feng, R., Du, B., Yan, L. & Wei, Q. (2011). Adsorption of benzoic acid from aqueous solution by three kinds of modified bentonites, *Journal of Colloid and Interface Science*, *359*, 499–504.

Xu, L. & Zhu, L. (2007). Structures of hexamethonium exchanged bentonite and the sorption characteristics for phenol, *Colloids and Surfaces A: Physicochem. Eng. Aspects*, *307*, 1–6.

Xu, L. & Zhu, L. (2009). Structures of OTMA- and DODMA-bentonite and their sorption characteristics towards organic compounds, *Journal of Colloid and Interface Science*, *331*, 8–14.

Yildiz, N., Gönülşen, R., Koyuncu, H. & Çalimli, A. (2005). Adsorption of benzoic acid and hydroquinone by organically modified bentonites, *Colloids and Surfaces A: Physicochem. Eng. Aspects*, *260*, 87–94.

Yu, W. H., Li, N., Tong, D. S., Zhou, C. H., Lin, C. X. C. & Xu, C. Y. (2013). Adsorption of proteins and nucleic acids on clay minerals and their interactions: A review, *Applied Clay Science*, 80–81, 443–452.

Zaghouane-Boudiaf, H. & Boutahala, M. (2011). Adsorption of 2, 4, 5-trichlorophenol by organo-montmorillonites from aqueous solutions: Kinetics and equilibrium studies, *Chemical Engineering Journal*, *170*, 120–126.

Zha, S. X., Zhou, Y., Jin, X.& Chen, Z. (2013). The removal of amoxicillin from wastewater using organobentonite, *Journal of Environmental Management*, *129*, 569-576.

Zhou, C. H., Zhang, D., Tong, D. S., Wu, L. M., Yu, W. H.& Ismadji, S. (2012). Paper-like composites of cellulose acetate-organo-montmorillonite for removal of hazardous anionic dye in water, *Chemical Engineering Journal*, *209*, 223–234.

Zhu, L. & Zhu, R. (2008). Surface structure of $CTMA^+$ modified bentonite and their sorptive characteristics towards organic compounds, *Colloids and Surfaces A: Physicochem. Eng. Aspects,* 320, 19–24.

Zohra, B., Aicha, K., Fatima, S., Nourredine, B. & Zoubir, D. (2008). Adsorption of Direct Red 2 on bentonite modified by cetyltrimethylammonium bromide, *Chemical Engineering Journal*, *136*, 295-305.

In: Bentonite
Editor: Ajay Kumar Mishra

ISBN: 978-1-63482-142-1

Chapter 4

BENTONITE AND ITS COMPOSITES: REMOVAL OF HEAVY METAL IONS FROM WATER

***Sadanand Pandey*[1,*]*, Ajay K. Mishra*[2] *and Shivani B. Mishra*[1,*]**

[1]Department of Applied Chemistry, University of Johannesburg, Johannesburg, South Africa

[2]Nanotechnology and Water Sustainability Research Unit, College of Science, Engineering and Technology, University of South Africa, Florida Campus, Johannesburg, South Africa

ABSTRACT

Bentonite extremely abundant low–cost, non toxic, eco friendly natural clay has been considered as a potential absorbent for removing pollutants from water and wastewater. It application is confined due to small surface area and presence of net negative surface charge, leading to its low adsorption capacity. Over the recent years, research on the development and characterization of Bentonite and its composites has received special attention because of it important properties to increase their adsorbent capacity, in order to remove different types of contaminants from drinking water are in great progress. This chapter start with the water pollution, causes of pollution, short description of different types of contaminants use for polluting water, role of nanotechnology, different advance methods use for water treatment were describe in this chapter. Apart from this, reviews of recent development of different types of Bentonite and its composite, and application for treating drinking water and their sources are the main beauty of this chapter. This chapter describes the versatile nature of Bentonite and its composite and their ability to absorb heavy metal ions, which are present in the drinking water. An attempt is also being made to include sufficient information like the type of polymer used for modifying the bentonite, their adsorption capacity and variables affecting the adsorption, so that complete information is available to the readers.

* Corresponding Author: E.mail: sadanand.au@gmail.com, smishra@uj.ac.za

1. Introduction

Water supports all forms of life on earth. The availability of clean water to the human population is of supreme importance. As world population increases, the consumption of water will also be increased. According to the surveyor on water, in the year 2025, half the population of this planet will be going to face the water-based infirmity or a water crisis. [1-2] This water crisis will also have a great impact on the biodiversity including life of flora and fauna.

Water pollution is the major problem in the global surrounding. And it is major causes of illness and mortality worldwide. [3] Now the awareness for the disposal of industrial wastewater, pretreatment and proper purification of water is the prime responsibility of all the people around the globe. Many laws, policies and a lot of technological advancement have been made related to recycling of industrial water and/or its treatment before it is discharged into the environment.

People are responsible for causing the water pollution on this earth. This water pollution is the result of using herbicides, pesticides, fertilizers, hazardous industrial chemicals, detergents, soaps, pathogens, heavy metals, organic solvents and textile dyes that find their way into water supplies (Table 1).

Table 1. Examples of water pollutants

EXAMPLES OF WATER POLLUTANTS	INCLUDES
Sewage and Domestic wastes	➢ Sewage includes human excreta, soap, detergent, metals, glass, plastic etc. ➢ Domestic sewage contains trace quantities of toxic metals such as Cu, Cr, Zn, Mn, Pb and Ni.
Industrial effluents	➢ It contains toxic chemicals, hazardous compounds, phenols, ketones, aldehydes, amines, oils, greases, dyes, suspended solids, non biodegradable matter, radioactive wastes etc.
Soap and detergents	➢ Soaps contain sodium or potassium salt of higher fatty acids ➢ Detergents contain ingredients like surfactants, builders, additives and stabilizers. The surfactants in detergents are alkyl benzene Sulphonate (ABS) and builders are usually a sodium phosphate. ➢ Most of the ingredients by soap are biodegradable. But that of detergents are not biodegradable. Hence detergents cause more water pollution than soap.
Pesticides	➢ It includes fungicides, bactericides, insecticides, nematicides, herbicides, rodenticides etc. ➢ Pesticides enter the aquatic system either directly or indirectly through the following sources: Rain water, Spray Drift, Run off from agricultural fields, industrial effluents and domestic sewage.
Fertilizers	➢ Excess of fertilizers not consumed by crops are washed away from land by rain water and pollute water
Heavy metals	➢ The heavy and toxic metals such as Cr, As, Hg, Cd, Pb, As, Co, Mn, and Fe are added from industrial process, domestic sewage, street dust, fossil fuel burning.

Types of Pollutants

There are major three different types of pollutants in wastewater. They are as follows:

✓ Organic water pollutants which include:

- Detergents
- Disinfection by-products such as chloroform
- Food processing waste, which can include oxygen-demanding substances, fats and grease
- Insecticides and herbicides, a large range of organohalides and other chemical compounds
- Petroleum hydrocarbons, including fuels and lubricants and fuel combustion byproducts.
- Tree and bush debris.
- Volatile organic compounds (VOCs)
- Chlorinated solvents such as Polychlorinated biphenyl (PCBs), Trichloroethylene
- Perchlorate
- Various chemical compounds present in personal hygiene and cosmetic products
- Drug pollution.

✓ Inorganic water pollutants which include:

- Acidity which is caused by industrial discharges (especially sulfur dioxide from power plants)
- Ammonia from food processing waste
- Chemical waste as industrial by-products
- Fertilizers containing nutrients--nitrates and phosphates
- Heavy metals from motor vehicles and acid mine drainage
- Silt (sediment) in runoff from construction sites, and burn practices or land clearing sites.

✓ Biological (pathogens) water pollutants which include:

- Coliform bacteria,
- Burkholderia pseudomallei,
- Cryptosporidium parvum,
- Giardia lamblia, Salmonella,
- Novovirus and other viruses,
- Parasitic worms (helminths),
- Pathogens from inadequately treated sewage discharges

✓ Recent human-caused water pollution

- Allegheny Energy, in 5 states including New York and New Jersey, in coal-fired power plant, has dumped tens of thousands of gallons of wastewater containing chemicals from the scrubbing process (to reduce air pollution) into the Monongahela River, which provides drinking water to 350,000 people and flows into Pittsburgh, 40 miles to the north. [4]
- One of the shocking environmental disasters hit China's waterways in 2013, An estimated 100,000 kilograms of dead fish have surfaced in Wuhan's Fu river. The thick floating blanket of rotting fish was poisoned by a recent chemical discharge. [5]
- The Elk River chemical spill occurred on January 9, 2014 when crude 4-methylcyclohexanemethanol (MCHM) was released from a Freedom Industries facility into the Elk River, a tributary of the Kanawha River, in Charleston in the U.S. state of West Virginia. Due to the spill, up to 300,000 residents within nine counties in the Charleston, West Virginia metropolitan area were without access to potable water. [6]
- Oil spills in the Gulf of Mexico (USA) and Yellow River (China) and the toxic red sludge flood 'in Hungary results in water pollution. [7-9]

1.1. Hazardous Effect of Heavy Metals and Organic Pollutants

Heavy metals such as Mercury (Hg^{2+}), Arsenic (As^{3+}), Chromium (Cr^{6+}), cadmium (Cd^{2+}) and lead (Pb^{2+}) are some of the most toxic inorganic contaminants found in the environment. Out of which cadmium (Cd^{2+}) and lead (Pb^{2+}) are very common one. These enter water through industries such as metal plating, smelting, mining, phosphate fertilizers, cadmium-nickel and lead batteries, textiles and alloy industries, paint industries, leather tanning, pigments, as well as from sewage. [10-12]. The harmful effect of some of the heavy metals are provided in Table 2.

Table 2. The harmful effect of some heavy metals

PURE CHEMICAL POLLUTANTS	HARMFUL AFFECT
• Cadmium	➢ Exposure to Cd^{2+} leads to renal damage [13] and osteoporosis. [14] Cadmium exposure also causes prostate and renal cancer [14,15].The permissible limit of Cd^{2+} in drinking water set by the WHO is 3 μg/L
• Lead	➢ Lead (Pb^{2+}) is found to be very hazardous heavy metal ion for the human being. The accumulation of Pb^{2+} in the body disturbs processes such as haemoglobin synthesis, renal function and cause neurological and behavioural disturbances in children. [12] ➢ The permissible limit of Pb^{2+} in drinking water set by the WHO is 10 μg/L.
• Chromium	➢ The major sources of chromium (3+) and (6+) in the air, soil and water are through combustion of fossil fuels and from domestic and industrial

PURE CHEMICAL POLLUTANTS	HARMFUL AFFECT
	waste. ➢ Respiration in chromium (VI) for a short period of time causes irritation of the nasal, cough and chest pain whereas swallowing results in abdominal pain, diarrhoea, heart failure, damaging the gut, liver and kidneys. Contact of Cr (6+) causes dermatitis or skin ulcers. ➢ EPA has an enforceable maximum contaminant level of total chromium in drinking water of 100 μg/L (100 ppb). [16]
• Arsenic	➢ Inorganic forms of arsenic are found in well water. It was found to be very much toxic than organic forms found in crustacean seafood [17]. ➢ Intake of large amount of arsenic can lead to acute intoxication results in the gastrointestinal, dermal, nervous, renal, hepatic, hematopoietic, cardiovascular, respiratory and ophthalmic systems [18, 19]. Chronic exposure from drinking water, coal burning or industrial sources can cause dermal lesions, neurological defects, atherosclerosis and cancer
• Zinc	➢ Zinc is a trace element that is essential for human health. However, too much zinc can cause eminent health problems, such as stomach cramps, skin irritations, vomiting, nausea and anemia. [20]
• Copper	➢ Excessive ingestion of copper brings about serious toxicological concerns, such as vomiting, cramps, convulsions, or even death. [21]
• Nickel	➢ Excessive intake of nickel may cause serious lung and kidney problems aside from gastrointestinal distress, pulmonary fibrosis and skin dermatitis. [22]

Organic Contaminants

Organic contaminants include pesticides, herbicides, decayed plant and animal tissues, gasoline and plasticizers (made from petroleum products). Plasticizers leach out of plumbing lines, fittings and storage tanks. Apart from this other organic contaminants like chlorinated organic solvents have been used in industries such as dyes and paints, adhesives, aerosols, lubricants etc. [23] This type of contaminants are known as human carcinogens, causes wide impact on public health.

Textiles industries are the major source of dye. Discharge of dye-containing effluents into the water environment is very hazardous because many of dyes released and their breakdown products are toxic, carcinogenic or mutagenic to life forms mainly because of carcinogens, such as benzidine, naphthalene and other aromatic compounds. [24, 25]

Food dyes like tartrazine were found to have an adverse effect on the behavior of young mice, inflamed the stomach lining and have adverse effects on the male reproductive system. [26] Neutral red was reported to bind to DNA [27, 28] and such interaction may result in mutations in DNA during replication. [29, 30]

2. Nanotechnology in Water Purification

Nanotechnology is the engineering and the art of manipulating matter at the nanoscale (1-100 nm). Nanotechnology has been extensively use in areas such as drug research, material chemistry and the environment, information and communication, heavy industry, textiles and cosmetics and energy. Nanotechnology-based methods for treatment of surface water, and wastewater contaminated by toxic metal ions, organic and inorganic solutes and microorganisms are an area that is now being widely explored.

Some of the examples of nanomaterial derived adsorbents for water treatment are:

- ✓ Cyclodextrin nanoporous polymer is reported to be effective in the removal of a range of organic contaminants [31, 32],
- ✓ Nanocatalyst-based technologies such as nanoscale zero-valent [33,34],
- ✓ Nanoscale titanium dioxide photocatalyst [35,36],
- ✓ Titanium dioxide nanoparticle adsorbent [37],
- ✓ Nanostructured iron oxide adsorbent [38]
- ✓ Magnetic nanoparticles such as magnetoferritin [39] are also being developed for use in water purification.

2.1. Polymer Nanocomposites

Polymer nanocomposites (PNC) consist of a polymeric material (e.g., thermoplastics, thermosets, or elastomers) and a reinforcing nanoscale material (nanoparticle). The nanoparticle has at least one dimension in nanometer scale. Polymer nanocomposites (PNC) show some significant major improvements such as:

- ✓ mechanical properties,
- ✓ gas barrier properties,
- ✓ thermal stability,
- ✓ fire retardancy,
- ✓ High selectivity
- ✓ High reactivity
- ✓ Regenerability
- ✓ Improved hydraulic properties
- ✓ Chemical resistance
- ✓ Robustness
- ✓ Low cost

There are many factors that affect the polymer nanocomposite properties:

- ✓ Synthesis methods such as, melt compounding, solvent blending, in-situ polymerization, and emulsion polymerization.
- ✓ Polymer nanocomposite morphology
- ✓ Types of nanoparticles and their surface treatments
- ✓ Polymer matrix such as crystallinity, molecular weight, polymer chemistry, and whether thermoplastic or thermosetting.

2.2. Types of Nanoparticles (Fillers)

There are different types of commercially available nanoparticles (filler) that can be incorporated into the polymer matrix to form polymer nanocomposites. The most commonly used nanoparticles in the literature are:

- ✓ Montmorillonite organoclays (Clay)
- ✓ Carbon nanofibers
- ✓ Polyhedral oligomeric silsesquioxane
- ✓ Carbon nanotubes
- ✓ Nanosilica
- ✓ Nanoaluminum oxide
- ✓ Nanotitanium oxide
- ✓ And many others.

2.2.1. Clay

Clay is one of the most abundant natural and inexpensive inorganic filler materials. Clay was introduced in the nanotechnology field as a new type of filler to produce polymer-clay nanocomposites (PCNs). [40-45] Depending on the ordering and degree of clay dispersion in a polymer matrix there are three types of nanocomposites that can be distinguished. [45] The structure of three different types are provided in Figure1.

- *Conventional composite* where the clay acts as a conventional filler
- *Intercalated nanocomposite* consisting of a regular insertion of the polymer in between the clay layers and
- *Delaminated (or exfoliated) nanocomposite* where 1 nm-thick layer are dispersed in the matrix forming a monolithic structure on the microscale.

Above all the three types of nanocomposites, delaminated nanocomposites are found to be of great importance because it maximises the polymer-clay interactions, making the entire surface of the layers available for the polymer. This should lead to most dramatic changes in mechanical and physical properties.

The preparation of polymer clay nanocomposites requires a good compatibility between the clay surface and the polymers or monomers. [41, 44, 45] The swellable clays, such as montmorillonite (MMT) are hydrophilic and therefore incompatible with hydrophobic polymers or monomers. [46, 47] The hydrophilicity of clay can be changed by surface modification. [44, 47]

The surface modification of clay can be typically performed by ion exchange of surface inorganic cations (e.g., Na^+, K^+, Ca^{2+}) by organic cationic surfactants. Polymer and oligomers with quaternary ammonium have been also used to modify the clay surface to obtain polymer-modified clay. [44, 45, 48]

Here in the present chapter we are focusing only on the polymer clay Nanocomposite. Nanoclay (Bentonite) is the most widely investigated nanoparticle in a variety of different polymer matrices for a spectrum of applications. According to geologists, bentonite is defined as "Absorbent aluminium phyllosilicate , essentially a rock formed of highly colloidal and plastic clays composed mainly of montmorillonite, a clay mineral of the smectite group, and is produced by *in situ* devitrification of volcanic ash. [49] The transformation of ash to bentonite apparently takes place only in water (certainly seawater, probably alkaline lakes, and possibly other fresh water) during or after deposition. [50, 51]

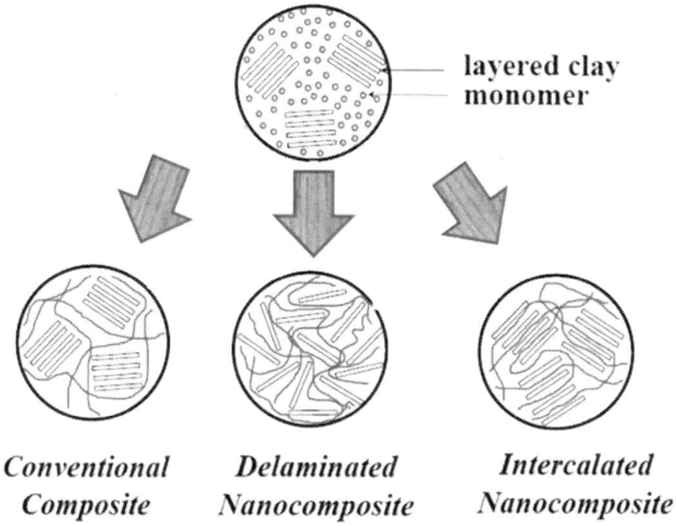

Figure 1. Structures of polymer clay nanocomposites.

Bentonite and Montmorillonite clays often interchange names and have similar properties yet vary from deposit to deposit. Actually it is not a significant difference. They are both Smectites. Most Smectite clays have Montmorillonite properties. The Montmorillonite name was actually given to Smectite clay that was found in the Montmorillon area of France. Clays are often named after their location. Bentonite was originally named for Smectite clay found near Fort Benton (Wyoming, USA). Most Smectites are now referred to as Bentonites or Montmorillonites. In addition to montmorillonite, bentonite may also contain feldspar, biotite, kaolinite, illite, cristobalite, pyroxene, zircon, and crystalline quartz. The absorbent clay was given the name bentonite by Wilbur C. Knight in 1898, after the Cretaceous Benton Shale near Rock River, Wyoming. [52] By extension, the term bentonite is applied commercially to any plastic, colloidal, and swelling clay regardless of its geological origin. Such clays are ordinarily composed largely of minerals of the montmorillonite group.

The term "montmorillonite" is also ambiguous and is used both for a group of related clay minerals and for a specific member of that group. [53] For the former use, smectite is more appropriate (a group of clay minerals that includes montmorillonite, saponite, sauconite, beidellite, nontronite, etc.) The Classification of silicates was performed by Bailey in 1980; [54], and Rieder et al., 1998. [55] The most-used clays for nanocomposites synthesis are smectite group which refers to a family of non-metallic clays primarily composed of hydrated sodium calcium aluminium silicate, with general formula of $(Ca,Na,H)(Al,Mg,Fe,Zn)_2$ $(Si,Al)_4O_{10}(OH)_2$ nH_2O. Smectite is a clay mineral having a 2:1 expanding crystal lattice (Figure 2). Members of the smectite group include the dioctahedral minerals montmorillonite, beidellite, nontronite, bentonite, and the trioctahedral minerals hectorite (Li-rich), saponite (Mgrich) and sauconite (Zn-rich). The basic structural unit is a layer consisting of two inward pointing tetrahedral sheets with a central alumina octahedral sheet. The layers are continuous in the length and width directions, but the bond between layers are weak and have excellent cleavage, allowing water and other molecules to enter between the layers causing expansion in the third direction. [45, 56, 57]

In the inner blocks, all corners of silica tetrahedra are connected to adjacent blocks, but in outer blocks some of the corners contain Si atoms bound to hydroxyls (Si-OH). These silanol groups at the external surface of the silicate, are usually accessible to organic species, and act as neutral adsorption sites. In addition to, some isomorphic substitutions occur in the tetrahedral sheet of the lattice of the mineral form leading to negatively charged adsorption sites which are occupied by exchangeable cations. [57] These characteristic make them powerful absorbents for organic molecules and organic cations. In order to improve the adsorption properties of clay absorbents also for organic anions, clay surface can be modified. There are many exchangeable cations on the clay surface therefore the cationic surfactants are generally used as modifiers. The characteristics of these so-called organoclays can be changed by variation of surfactant properties, such as alkyl chain length, etc. The surface properties of the clays modified by surfactants alter from organophobic to organophilic, which aids in improving clay adsorption capacities for organic compounds. [59] While crude clay minerals are effective for the adsorption of cations, organo-modified clays may adsorb negative and hydrophobic molecules. [57]

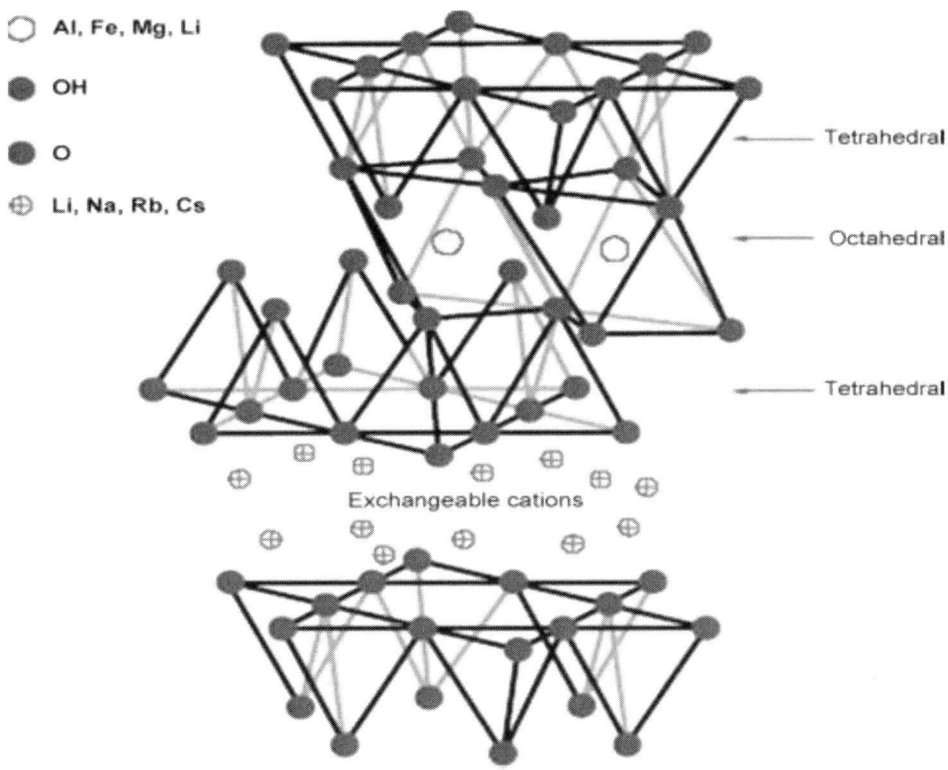

Figure 2. The structure of a 2:1 layered silicate (Beyer, 2002). [56] [Adopted from Beyer by copyright permission 2014 of Elsevier Science Ltd., UK].

Bentonite and Montmorillonite have similar mineral properties. The most important of which are attributed to the Smectite Family of clays. The Cation Exchange Capacity CEC and the Specific Surface area of Smectites are considerable larger than other families of clay. There absorption capacity is as much as 8 times greater than other clays. [60] Polymer composites prepared with MMT (or Bentonite) is commonly referred to as *montmorillonite composites* (MMC), Bentonite composite or *polymer clay nanocomposites* (PCNC). Of the smectite family of clays, montmorillonite has received much attention owing to its high surface area, high cation exchange capacity, swelling properties in water and its potential for ion-exchange. [61]

In water treatment, polymer-clay nanocomposites have been used in the preparation of nanofiltration membranes [62] and for adsorption. [63] Montarges et al. [64] also reported synthesizing a nanocomposite of poly (ethylene oxide)-poly (propylene oxide)-Al 13-pillared montmorillonite for possible application in the removal of chlorinated organic from water by adsorption. Before we start the review of polymer clay nanocomposites use for removal of contaminants from water. We will try to review the different methods use for treatment of water.

3. Different Water Treatment Methods

As we have mention earlier in this chapter, that there are three major types of pollutants in wastewater: Organic, Inorganic and biological pollutants. All these types of pollutants are harmful for human being. So there is urgent need for removing these important and dangerous pollutants from the wastewater. There are several methods available for the removal of these pollutants from wastewater.

3.1. Chemical Precipitation

Chemical precipitation is a conventional technology used to treat mining-influenced water (MIW), including acid mine drainage, neutral drainage, and pit lake water. Chemical precipitation processes involve the addition of chemical reagents, followed by the separation of the precipitated solids from the cleaned water. Chemical precipitation is effective and the most widely used process in industry [65] because it is relatively simple and inexpensive to operate. Precipitation can be induced by the addition of an alkali, sulfide, coagulant, or other reagent that will bond with dissolved metal ions. Thus conventional chemical precipitation processes include

- ✓ Hydroxide precipitation [66-68]

Precipitation can be induced by the addition of an alkali sources which include caustic sodium hydroxide (NaOH), hydrated lime (Ca $(OH)_2$), quick lime (CaO), limestone ($CaCO_3$), and magnesium hydroxide ($Mg(OH)_2$).

Hydroxide precipitation process using $Ca(OH)_2$ and NaOH in removing Cu^{2+} and Cr^{6+} ions from wastewater was evaluated by Mirbagheri and Hosseini. [67]

Charerntanyarak [68] employed chemical coagulation and precipitation by lime to treat synthetic wastewater consisting of Zn, Cd, Mn and Mg at the concentration of 450, 150, 1085 and 3154 mg/L, respectively.

✓ Sulfide Precipitation. Bhattacharyya et al. & Özverdi and Erdem [69,70] reported Sulfide reagents used to cause precipitation of contaminants include iron sulfide (FeS), sodium hydrosulfide (NaHS), sodium sulfide (Na2S), calcium sulfide (CaS), and biogenic sulfide generated in situ by sulfate reduction.

Sulfide precipitation with Na_2S is found to be highly effective to obtain a high degree of separation of heavy metal cations (Cd, Zn, Cu, and Pb) and of the oxyanions of arsenic and selenium from complex wastewaters. The removals of Cd, Zn, and Cu from the actual wastewaters are greater than 99%, and As and Se removals are 98 and >92%, respectively. Cd, Cu, and Zn concentrations in the range of 0.05 to 0.1 mg/1 can be achieved with sulfide precipitation. [69]

Özverdi and Erdem [70] investigated pyrite and synthetic iron sulfide to remove Cu^{2+}, Cd^{2+} and Pb^{2+}.

The Chemical precipitation has found to have following advantages:

- Permanent
- Immediate results
- Efficient
- Easily implemented
- And easy to monitor

The Disadvantages or limitations of chemical precipitation include high cost

- Not applicable for all cases
- Requires operation and maintenance (O&M)
- Requires power
- May generate a waste product

✓ Heavy metal chelating precipitation

As, we have already seen the above that conventional chemical precipitation processes have many limitations. Thus now as an alternative, many companies use chelating precipitants to precipitate heavy metals from aqueous systems. For examples;

- The effectiveness of commercial heavy metal precipitants, trimercaptotriazine, potassium/sodiumthiocarbonate and sodiumdimethyldithiocarbamate was examine and reviewed by Matlock et al. [71a]
- A new thiol- based compound, 1,3-benzenediamidoethanethiol ($BDET^{2-}$) dianion was designed by Matlock et al. [71b,c]. $BDET^{2-}$ is found to precipitate mercury in the leachate solution and heavy metals from acid mine drainage.

- Two dithiocarbamate-type supramolecular heavy metal precipitants, N,N'-bis-(dithiocarboxy)piperazine(BDP) and 1,3,5 hexahydrotriazinedithiocarbamate (HTDC) in treating complex heavy metal wastewater was designed by Fu et al. [72,73]
- Ying & Xu [74] investigated the chemistry of complexes of copper, nickel and chromium with EDTA, ammonia and citric acid under co-exist of dithiocarbamate chelating agent.
- Xu et al. [75] reported the removal of Cd^{2+}, they have develop Potassium butyl dithiophosphate (PBD) as a new chelating agent. The Cd^{2+} removal rate was found to be 99%. If Cd^{2+} co-exists with Pb^{2+} and Cu^{2+}, the affinity of the chelating agent with these three heavy metal ions found the following order : $Cu^{2+} > Pb^{2+} > Cd^{2+}$.

3.2. Ion Exchange

Ion-exchange processes have been widely used to remove heavy metals from wastewater due to some important advantages, such as high treatment capacity, high removal efficiency and fast kinetics. [76] Typical ion exchangers are ion exchange resins (functionalized porous or gel polymer), zeolites, *montmorillonite*, clay, and soil humus. Ion exchangers are either cation exchangers that exchange positively charged ions (*cations*) or anion exchangers that exchange negatively charged ions (anions). There are also *amphoteric* exchangers that are able to exchange both cations and anions simultaneously. Ion-exchange processes are used to separate and purify metals, including separating uranium from plutonium and other actinides, including thorium, and lanthanum, neodymium, ytterbium, samarium, lutetium, from each other and the other lanthanides Ionic charge also plays an important role in ion-exchange process.

Kang et al.[76] obtained the Influence of ionic charge on removal of Co^{2+}, Ni^{2+} and Cr^{3+} on an Amberlite IRN-77 cation- exchange resin.

Abo-Farha et al. [77] tested the influence of ionic charge on the removal of Ce^{4+}, $Fe3^{+}$ and Pb^{2+} from aqueous systems by cation-exchange resin purolite C100. Abo-Farha et al. (2009) found that the metal ions adsorption sequence can be given as $Ce^{4+}>Fe^{3+}>Pb^{2+}$.

Scientists like Motsi et al., Ostroski et al., and Taffarel and Rubio [78-80] have demonstrated that zeolites exhibit good cation-exchange capacities for heavy metal ions under different experimental conditions.

Doula and Dimirkou and Doula [81,82] reported that the surface of clinoptilolite loaded with amorphous Fe-oxide species would significantly enhance the exchange capacity of clinoptilolite.

Doula [82] shows the removal of Cu, Mn and Zn from drinking water by using clinoptilolite-Fe system.

Bayer [83] recommended selective removal of metal ions, especially Pb(II) by chelating ion exchanger with functional iminodiacetate groups Lewatit TP-207. The removal of same metal ion i.e., Pb(II) by combining of cation exchange with precipitation was performed by Lee and Hong. [84] Pb(II) removal from wastewaters by Italian zeolites, like chabazite and phillipsite was performed by Colella [85], Pansinini et al. [86] and Torracca et al. [87]

Gardiner and Munoz, De Jong and Rekers, and Ocenasek et al. used various types of ion exchangers e.g., strongly acidic cation exchangers, weakly and strongly basic anion exchangers as well as on selective ion exchangers for Hg (II) removal. [88-90]

Bolto and Pawłowski, Warshawsky, and Dorfner used N and S functional groups ion exchangers—Varion BTAM (Nike), thiocarbamate group—Varion BTKM (Nike), mercapthile group- Varion BSM (Nike) and thiourea functional groups for selective removal of Hg(II). [91-93]

Kauspediene et al., design the Polyacrylate carboxylic cation exchangers KB-2 and KB-2-12P were used for simultaneous removal of both Cd(II) ions and non-ionic surfactants from aqueous solutions. [94]

Two natural zeolites- phillipsite and chabazite for Cd (II) ions removal from wastewater was studied by Colella et al. [85] Removal of selective Ni^{2+} from nickel plating by macroporous carboxylic cation exchanger Wofatit CA- 20 in the sodium form was performed by Halle et al. [95] Untea et al. shows the high selectivity for Cr(VI) ions by Macroporous strongly basic anion exchanger— Lewatit MP-500A. [96] Kurashvili et al. also shows the high distribution coefficient of Cr (VI) ions removal by applying anion exchange fibres with pyridine groups. [97]

3.3. Biosorption Processes

Biosorption can be defined as the ability of biological materials to accumulate heavy metals from wastewater through metabolically mediated or physico-chemical pathways of uptake. [98] Typical biosorbents can be derived from three sources as follows [99]:

- ➢ Non-living biomass such as bark, lignin, shrimp, krill, squid, crab shell, etc.
- ➢ Algal biomass
- ➢ Microbial biomass, e.g., bacteria, fungi and yeast.

The major advantages of biosorption over conventional treatment methods include [100]:

- ✓ Low cost;
- ✓ High efficiency
- ✓ Minimisation of chemical or biological sludge
- ✓ No additional nutrient requirement
- ✓ Regeneration of biosorbent; and
- ✓ Possibility of metal recovery.

Biosorption processes are particularly useful for the treatment of dilute heavy metal wastewater. Limitation, the separation of biosorbents would be difficult after adsorption. Because of which it cannot be recycle and reuse further.

Some most common inexpensive types of biosorption are, non-living plant material such as potato peels [101] sawdust [102], black gram husk [103], eggshell [104], seed shells [105], coffee husks [106], sugar-beet pectin gels [107]and citrus peels [108], etc.

Some other recently used biosorption are as follow; Biosorption of $Cu^{2+}$and Zn^{2+} using dried marine green macroalga Chaetomorpha linum [109] the biosorption of Cu^{2+} ,Cd^{2+} ,Pb^{2+}, and Zn^{2+} using dried marine green macroalga Caulerpa lentillifera [110] the biosorption of chromium from waste- water using green alga Ulva lactuca [111] and the biosorption of lead (II) from wastewater by green algae Cladophora fascicularis. [112] The biosorption of Cu2+ and Zn2+ by dried marine green macroalga (C. linum) was investigated by Ajjabi and Chouba. Biosorption of heavy metals in aqueous solutions by bacteria includes Bacillus cereus [113], Escherichia coli [114], [115] Pseudomonas aeruginosa [116], Bacillus fir mus [117], Corynebacterium glutamicum [118], Pseudomonas aeruginosa [119], Pseudomonas putida [120], Streptomyces rimosus [121], etc. Table 3 summarizes some of the important results of metal biosorption using bacterial biomasses and fungi along with their adsorption capacity.

Table 3. Some important results of metal biosorption using bacterial biomasses and fungi along with adsorption capacity (mg/g)

Metal ions	Biomass	Biosorption capacity	References
Bacteria species			
Pb	*Bacillus firmus*	467	[117]
	Corynebacterium glutamicum	567.7	[118]
	Pseudomonas aeruginosa	79.5	[119]
	Pseudomonas putida	270.4	[120]
	Streptomyces rimosus	135	[121]
Zn	*Bacillus firmus*	418	[117]
	Aphanothece halophytica	133	[122]
	Pseudomonas putida	17.7	[123]
	Streptomyces rimosus	80	[124]
	Thiobacillus ferrooxidans	172.4	[125]
Cu	*Bacillus firmus*	381	[117]
	Bacillus subtilis	20.8	[126]
	Enterobacter Sp.	32.5	[127]
	Pseudomonas cepacia	65.3	[128]
	Pseudomonas putida	96.9	[129]
	Streptomyces coelicolor	66.7	[130]
Cd	*Aeromonas caviae*	155.3	[131]
	Enterobacter Sp.	46.2	[127]
	Pseudomonas Sp.	278	[132]
Fe(III)	*Streptomyces rimosus*	122	[121]
Cr(VI)	*Aeromonas caviae*	284.4	[131]
Ni	*Bacillus thuringiensis*	45.9	[133]
Pd	*Desulfovibrio vulgaris*	106.3	[134]
Pt	*Desulfovibrio fructosivorans*	32.3	[134]
U	*Bacillus megaterium*	37.8	[135]
Th	*Micrococcus luteus*	77	[135]
Penicillium species.			
Cd	*Penicillium canescens*	102.7	[136]
As(III)	*Penicillium canescens*	26.4	[136]
Pb	*Penicillium chrysogenum*	116	[137]
Cd	*Penicillium chrysogenum* (modified)	210.2	[138]
U	*Penicillium chrysogenum*	70	[139]
Pb	*Aspergillus niger* (pretreated with NaOH)	28.9	[140a]
Cu	*Aspergillus niger*	9.53	[140b]
Tc, U, Cs, Eu,	*Aspergillus niger*	-	[141]

3.4. Membrane Filtration

Membrane filtration technologies with different types of membranes show great promise for heavy metal removal for their high efficiency, easy operation and space saving. The membrane processes used to remove metals from the wastewater are:

- Ultra-filtration (UF),
- Reverse osmosis(RO),
- Nano filtration (NF) and
- Electrodialysis (ED)

All types of membrane filtration processes are currently limited by the high cost incurred due to membrane fouling and replacement.

- *Ultrafiltration (UF)* is a variety of membrane filtration in which forces like pressure or concentration gradients leads to a separation through semipermeable membrane.Some examples of UF are; Zinc removal from synthetic wastewater by the micellar enhanced ultrafiltration using SDS was performed by Landaburu-Aguirre et al. (2009). [142] Removal of heavy metal ions such as Cd^{2+},C^{2+} ,Ni^{2+} ,Pb^{2+} and Zn^{2+} from synthetic water by the micellar enhanced ultrafiltration using two anionic surfactants: SDS and linear alkylbenzene sulfonate (LAS) in a lab-scale membrane system was investigated by Sampera et al. (2009). [143] Apart from this certain complexing agents such as polyacrylic acid (PAA), [144] polyethyleneimine (PEI) [145,146], diethylaminoethyl cellulose [147] and humic acid, [148] etc., have been proven to achieve selective separation and recovery of heavy metals with low energy requirements.
- ***Reverse Osmosis (RO)*** is a process in which heavy metals are separated by a semi-permeable membrane at a pressure greater than osmotic pressure caused by the dissolved solids in wastewater. RO is an increasingly popular wastewater treatment option in chemical and environmental engineering. Even nowdays in every home, this RO technology are used for cleaning the drinking water. By using RO process two important metal ions Cu^{2+} and Ni^{2+} are successfully removed and rejection efficiency of these ions enhance by 99.5% by using Na_2EDTA. [149]
- *Nano filtration (NF)* is a relatively recent membrane filtration process. It is intermediate process between UF and RO. It is use for purpose of softening (polyvalent cation removal) and removal of disinfection by-product precursors such as natural organic matter and synthetic organic matter (R. D. Letterman, 1999). [150] It is used for the rejection of heavy metal ions such as chromium [151], copper, [152] silver [153] arsenic [154] from wastewater. NF process benefits from ease of operation, reliability and comparatively low energy consumption as well as high efficiency of pollutant removal. [155]
- *Electrodialysis (ED)* is another membrane process for the separation of ions across charged membranes from one solution to another using an electric field as the driving force. The membranes uses during ED process are actually of two types: cation-exchange and anion-exchange membranes. This process has been widely used for the production of drinking water from brackish water and seawater, treatment of

industrial effluents, recovery of useful materials from effluents and salt production. [156] ED has also proven a promising method in heavy metal wastewater treatment. Investigation of separation of Cr^{3+} from sodium ion by using modified cation-exchange membranes and effect of operating parameters on Pb^{2+} separation from wastewater using ED was performed by Lambert et al. (2006) [157] and Mohammadi et al. (2004) [158] respectively.

Electrochemical treatment: This method involves the plating-out of metal ions on a cathode surface and can recover metals in the elemental metal state. Electrochemical wastewater technologies have certain limitations; involve relatively large capital investment and the expensive electricity supply, so they haven't been widely applied.

Electrochemical system along with with aluminium electrodes for removing Zn^{2+},Cu^{2+},Ni^{2+},Ag^{+} and $Cr_2O_7^{2-}$ was investigated by Heidmann and Calmano (2008) [159]. The removal of 1470ppm of Cr6+ by electrochemical method was performed by Ölmez (2009). [160]

Advanced oxidation process (AOT): It refers to a set of chemical treatment procedures designed to remove organic (and sometimes inorganic) materials in water and waste water by oxidation through reactions with hydroxyl radicals (·OH) (G.William, 1987). [161] AOT is chemical oxidation steps include the following processes: (1) ozonation, (2) combination of ozonation with UV-irradiation or hydrogen peroxide (peroxon), (3) combination of hydrogen peroxide with UV irradiation, (4) Fenton's reagent (Fe^{2+}/H_2O_2) and the photo-chemically enhanced Fenton process. But the costs of AOPs are too high, since a continuous input of expensive chemical reagents are required to maintain the operation of most AOPs system. Removal of natural organic matter from drinking water by AOT was performed by (Matilainen& Sillanpää, 2010). [162]

Ozonation: Ozonation is a chemical process that adds ozone (O_3) to a substance. Drinking water can be ozinatated to kill microorganisms that can cause waterborne diseases. Removal of natural organic matter (NOM) from drinking water supplies by ozone-biofiltration was studied by Raymond M et al., 1999. [163]

3.5. Adsorption

Adsorption is now recognized as an effective and economic method for wastewater treatment. Among all the different type of methods, removal of pollutants by adsorbents is found to be simple (in terms of operation), cost-effective and efficient methods for the removal of both organic and inorganic pollutants from contaminated waters. [164,165] In the last few decades, there are different types of adsorbents had been used for the removal of pollutants from wastewater.

But in the present chapters we are reviewing polymer-clay nanocomposites for wastewater treatment. Attention is given only to natural clay (Bentonite) for polymer clay nanocomposites synthesis and its application in water treatment. As we have discuss earlier that the polymer clay nanocomposites shows some of the significant properties. Because of enhancing these properties, different types of polymer are being used by different workers. Some recent polymer clay nanocomposites for improving the mechanical, thermal, chemical and opticals are reviewed here.

Shipeng et al., 2009 [166] investigated the influence of poly (ethylene glycol)/ montmorillonite hybrids on the rheological behaviors and mechanical properties of polypropylene. Young et al., 2009 [167] studied the effect of silane functionalization of montmorillonite on epoxy/montmorillonite nanocomposites which results in improve mechanical and dynamic mechanical properties. Zehetmeyer et al., 2013 [168] investigated the improvement in Morphological, optical, and barrier properties of PP/MMT nanocomposites. Pascual et al., 2009 [169] studied the Influence of the compatibilizer/ nanoclay ratio on final properties of polypropylene matrix modified with montmorillonite-based organoclay. Yanlian et al., 2012 [170] improved the UV-resistant property of lacquer film significantly by solution intercalation method by using raw lacquer/multihydroxyl polyacrylate/organophilic montmorillonite nanocomposites. Xi et al., 2011 [171] prepared and investigated exfoliated epoxy/clay nanocomposites for tensile modulus and tensile strength which was increased by 64.2 and 52.3%, respectively. Flexural modulus enhanced by 7.1%, flexural strength by 14.6% and notch impact strength by 37.7%. Even the glass transition temperature and thermal decomposition temperature moved to higher temperatures by 10.1 and 5.7 °C, respectively. Azhar et al., 2014 [172] investigated the improved thermal stability of polymer–montmorillonite nanocomposites based on natural polymers (chitosan, alginate, gelatin and starch) and montmorillonite (MMT) prepared by using solution intercalation method.

4. Polymer Clay Nanocomposites for Removal of Heavy Metals

Heavy metals such as Mercury (Hg^{2+}), Arsenic (As^{3+}), Chromium (Cr^{6+}), cadmium (Cd^{2+}) and lead (Pb^{2+}) are some of the most toxic inorganic contaminants found in the environment. These contaminants in water cause serious effects on the health of human beings, animals, and plants. There are many scientists and researchers are working for removal of heavy metal ions from wastewater. Application and efficiency of different type of natural clay and their composites in removing various metal ions are reviewed in the following sections.

Jiang & Zeng, 2003) modify clays, MMT K10 and KSF, by organocations (HDTMA), polymeric Al/Fe species and a complex of HDTMA-polymeric Al/Fe respectively. Adsorption behaviors of the resulting modified clays were evaluated using model waters containing either organic (phenol) or inorganic (Cu^{2+}) or both contaminants. It follows the Freundlich adsorption Isotherm of phenol and Cu onto modified clay and correlation coefficient was found to be 0.99. [173]

Cobalt adsorption efficiencies of kaolinite and MMT were studied by Angove et al. [175] and Bhattacharyya and Sen Gupta. [174] Angove et al. [175] found the Langmuir capacities as 1.5mg/g at 313 k and Bhattacharyya and Sen Gupta found it to 11.2mg/g for kaolinite. For MMT, Bhattacharyya and Sen Gupta found the Langmuir and freundilich values to be 28.6 and 4.6, respectively.

Guerra et al. 2013 studied the Pb^{2+} adsorption study of natural Brazilian bentonite (BBT) and modify natural BBT. Modification was performed by functionalization of 3-aminopropyltrietoxisilane (APS) and 3,2-aminoethylaminopropyltrimetoxisilane (AEAPS) on the surface of component minerals of BBT and designated as BBT_{APS} and BBT_{AEAPS}

respectively. The maximum Pb2+ adsorption capacities of bentonite types were 20.6843, 2 7.6524, and 29.5413mgg^{-1} for BBT, BBT_{APS} ,BBT_{AEAPS} respectively by using a Perkin Elmer Flame Atomic Absorption (AA) Spectrometer. The favorable optimized condition for Pb^{2+} binding was pH6 and adsorption time is 180min in all the three cases. The maximum binding was observed in BBT_{AEAPS}. Experimental results reveal the adsorption is langmuir adsorption Isotherm. With KL (dm^3 mmol^{-1}) values 0.4222, 0.5671, 0.6156, for BBT, BBT_{APS} ,BBT_{AEAPS} respectively; and R^2 values 0.9856, 0.9876, 0.9987, for BBT, BBT_{APS} ,BBT_{AEAPS} respectively. Experimental data was achieved by the pseudo-second order model, indicating that the chemisorption rate seems to be the controlling step in the procedure with ks (g mmol^{-1} min^{-1}) values 0.7761, 0.8145, and 0.919, for BBT, BBT_{APS} , and BBT_{AEAPS} respectively. The correlation coefficient of the pseudo-second-order rate equation (R^2) value 0.9999, 0.9998, 0.9997, for BBT, BBT_{APS} ,BBT_{AEAPS} respectively. The thermodynamic study was also shown in this article. It shows the increase in the temperature result in enhancing Pb(2+) removal efficiency by15.40%,13.37%,and11.22% for BBT_{APS}, BBT_{AEAPS}, and BBT, respectively, in 4h, at 413 K. The activation energy (Ea) KJmol^{-1}was found to be 25.51, 30.16 and 31.76 for BBT, BBT_{APS} and BBT_{AEAPS} respectively. [176]

Guerra et al. [176] have given the explanation for maximum binding on the bases of surface area. The BET analysis revealed that the surface area of BBT_{AEAPS} is 612.3m^2 g-^{1} has a significantly higher surface area than the BBT_{APS} (597.3) and BBT (34.1). It was also observed that BBT_{AEAPS} has pores <2 nm in diameter (micropore) in addition to some mesopores (44 nm diameter).The natural bentonite has small N_2 BET surface areas and only mesopores because N_2, as opposed to H_2O, OH^-, and H_3O^+, is unable to penetrate in to the octahedral position instructure of interstratified smectite–illite may have in the center a vacancy.

Triantafyllou et al. has Investigated Na- Activated bentonite for removal of Ni and Co from aqueous solution. [177]

Talaat et al., 2011 reported the raw and treated Egyptian kaolin as well as Ca and Na-bentonites (Ca-B and Na-B) for heavy metals removal. The treatment was performed using acid washing and calcination. The XRD pattern for Na-bentonite reveals that it contains 20.3 %, 15.5% kaolin, 33.9% quartz and 30.5% K-feldspar (as microcline) as compared to Egyptian bentonite of El-Fayoum area which contains 40-60% montmorillonite and 20-30 % kaolinite, while bentonite of Abu Tartour contains 5% kaolinite, 56% montmorillonite, 9% illite and 11% free quartz. Adsorption of Cr (III), Ni (II), Cd (II), Cu (II), Zn (II) and Pb (II) salts at low concentrations were investigated in batch mode. The clay dose was 0.5 g/l with metal concentrations from 0.3 mg/l to 10.3 mg/l. The maximum attained Cr (III) adsorption capacities for the tested clays are 9.2 mg/g, 14.7 and 15.7 mg/g for Kr, Ca-B and Na-B, respectively. As for Ni (II) adsorption, Ca-B shows best results reaching 11.6mg/g adsorption capacity. It is concluded that Ca-B had the highest adsorption capacities for Cu, Cd, Pb, Zn and Ni. It is also observed that the selectivity of metals adsorbed is in the order of Pb>Cr>Cd>Cu>Zn>Ni. This may be attributed to the affinity of combination of each metal on the clay. The adsorption of Cd (II), Cu (II) and Ni (II) were well represented using Freundlich isotherm, while that of Pb (II) and Zn (II) fitted well with Langmuir isotherm. The results indicate that the use of bentonite could provide a cheap and technically viable option for heavy metals removal from polluted water. [178]

Mohammed-Azizi et al. 2013 reported Algerian natural clay, a bentonite of Mostaganem (N.W. Algeria), has been acid-activated (HCl). The acid-activated clay (AAC) was employed as adsorbent for the removal of Zn^{2+}, Pb^{2+}, Cu^{2+}, and Ni^{2+} ions from aqueous solutions using adsorption method. The optimum parameters of our study obtained were pH values range (4–6), mass of AAC (1 g), shaking time (1 h), and initial concentration of metal ions (10mgL^{-1}). The results have shown that Algerian bentonite clay had significant potential for removing Zn^{2+}, Pb^{2+}, Cu^{2+}, and Ni^{2+} ions from polluted water. [179]

Kaya & Oren 2005 Investigated natural bentonite and Na- enriched bentonite as an adsorbent for Zn^{2+} removal at different pH values, initial metal ion concentrations and slurry concentrations. Results obtained from the effect of pH on the adsorption capacity of the bentonite indicated that the major factor which affected this parameter are competition of the H+ ions with Zn^{2+} ions (under pH 4), ion exchange (pH 4–7), and participation of the heavy metal species to the adsorption and precipitation of some onto the bentonite (pH 8). The cation exchange capacity (meq/100 g) and specific surface area (m^2/g) of natural bentonite is 393.44 & 132.33 respectively; and for Na-enriched bentonite is 441.96 & 67.11 respectively. [180]

Hefne, et al.2010 reported Ag+ could be adsorbed and thus higher significant amounts removed by Natural Bentonite from aqueous solutions. The removal % of Ag^+ increased with the increase of contact time, amount of adsorbent and pH. Langmuir model better represented the sorption process than the Freundlich model. The Q_{max} (mg/g) value was 53 and 69 at 293 & 333 respectively. The correlation coefficient of the Langmuir model equation (R^2) value 0.989, Kinetic modeling results showed that the pseudo second-order equation with K_2 (g/mg .min) is 0.05 and (R^2) value is 1. High adsorption capacity of washed bentonite (WB) compared to Natural bentonite (NB) and calcined bentonite (CB). Depending on the negative value of ΔG, the adsorption of Ag+ on NB surfaces was spontaneous and the adsorption was endothermic process. The value of ΔH (kJmol-1) and ΔS (kJmol-1K-1) is 4.15 and 0.05 respectively. [181]

Ayari et al., 2007 reported the amount of iron, aluminum and magnesium compounds decreased in the modified clay. Use of purified and modified bentonite clay by HCl (3M) for the removal of some heavy metals (Zn, Pb, Ni) from aqueous solutions has been investigated. They are the most impurities of industrials effluents. The clay fraction used in this work was an interstratified illit-smectite with 89 meq/100 g cation exchange capacity of calcined clay and 504 m2 g-1 specific surface areas, which is used in the purified and modified form, to test its proprieties of sorption for the removal of some toxic metals (Zn, Pb, Ni) from aqueous solutions. The sorption process is reflected by Langmuir and Freundlich-type isotherm. The order of the removal capacity is $Pb^{2+} > Zn^{2+} > Ni^{2+}$ for both natural and treated sorbents. Equilibrium modeling of the adsorption showed that the adsorption of these three metals was fitted to a Freundlich isotherm in all cases. [182]

Olu-Owolabi et al., reported Pre-modification of bentonite clay with goethite, humic acid, and a binary mixture of goethite and humic acid reagents increased its cation exchange capacity from 95 to 105.32, 120.4, and 125.8 meq/ 100 g of bentonite clay, respectively. The effective pre-modification of bentonite clay with goethite, humic acid, and goethite–humic acid reagents was confirmed from their Fourier transform infrared spectra which suggested that modification was effective on the AlAlOH and Si–O sites for goethite and humic acid modification and AlAlOH for goethite– humic acid modification. The goethite– humic acid-modified bentonite gave the best adsorption capacity of ≈10 and 16 mg/g at 30 and 50°C, respectively, for both metal ions. [183]

Bourlival et al., 2012 reported two natural Greek clays, bentonite from Milos Island and vermiculite from Askos area, North Greece were used for Ni (II) removal from aqueous solutions. Batch adsorption studies were carried out with various initial Ni (II) concentrations, adsorbent dosages, pH, and contact time. The adsorption takes place at pH 5. Overall the kinetic studies showed that the nickel adsorption process followed pseudo-second-order kinetics. The pseudo first-order plots of log (qe-qt) vs. t had good linearity R2 is 0.998 and the rate constant, k_2 was 0.58 x 10^{-2} g mg^{-1} min for bentonite and 0.52 x 10^{-2} g mg^{-1} min for vermiculite. The adsorption data gave good fits with Langmuir (R^2>0,95) and Freundlich (R^2>0,93) isotherms and yielded Langmuir monolayer capacities of 26.32 to 38.46 mg/g and Freundlich adsorption capacities of 2.95 to 3.05 mg(1−1/n) L1/n/g for the natural bentonite and vermiculite, respectively. [184]

Bekele1 et al., 2014 reported Natural bentonite clay has been activated by 2M HCl and the result obtained by FIT-IR spectroscopy characterization indicated the expected activation. Nitrate ions removal capacity of acid activated Ethiopian bentonite clay from aqueous solution was studied using batch experiment and the results obtained indicates that greater than 80% of nitrate ions has been removed from aqueous solution at optimum conditions of contact time (90 minutes), acid activated Ethiopian bentonite clay (2.0 g), initial nitrate concentration (250 mg/L), pH (5), and temperature (30°C). The adsorption data gave good fits with Langmuir (R^2= 0.9946) and Freundlich (R^2= 0.8874) isotherms and yielded Langmuir monolayer capacities of 8.68 and Freundlich adsorption capacities of 3.68 for acid activated bentonite. [185]

Al-Shahrani, 2014 study the removal of cobalt from wastewater using Saudi activated clay (bentonite). It was found that adsorption of cobalt ions on Saudi activated bentonite was relatively fast and the equilibrium was reached after 30 min. Adsorption was also dependent on solution pH where cobalt removal percentage gradually increased with increasing solution pH up to 99% at pH 8. Moreover, about 100% cobalt removal was observed when solution pH was increased to more than 8. The adsorption data gave good fits with Langmuir (R^2= 0.975) and Freundlich (R^2= 0.936) isotherms and yielded Langmuir monolayer capacities of 7.3 and Freundlich adsorption capacities of 2.33 for acid activated bentonite. Furthermore, the value of the mean free energy (E=91.3KJ/mole) determined with the D–R equation showed that adsorption process is chemisorptions. Cobalt adsorption into Saudi activated bentonite was well represented by the pseudo-second-order kinetic model with correlation coefficient was around 1.0. [186]

Bedelean et al., 2010 reported bentonite sample from the Petreşti deposit, Cluj County (Romania), was used to remove heavy-metal ions (Cd^{2+}, Pb^{2+}, Cr^{3+}) from wastewaters. The bentonite sample was used in powdered form (d < 0.2 mm) without any chemical treatment. The bentonite sample proved to be efficient for the removal of these heavy-metal ions; removal efficiencies up to 100% (lead removal) were achieved. [187]

Sandy et al., 2012 reported the adsorption of copper ions from an aqueous solution using linear alkylbenzene sulfonate (LABORATORIES) modified bentonite (organo-bentonite) are reported. Modification of bentonite was performed via microwave heating with a concentration of LABORATORIES surfactant equivalent to 1.5 times that of the cation exchange capacity (CEC) of the raw bentonite. Several adsorption equations (e.g., Langmuir, Freundlich, Sips and Toth) with temperature dependency were used to correlate the equilibrium data.. The Sips model had the best fit for the adsorption of copper ions onto organo-bentonite. For the kinetic data, the pseudo-second order model was superior to the pseudo-first order model. Thermodynamically, the adsorption of copper ions occurs via chemisorption and the process is endothermic (ΔH^{0}>0), irreversible (ΔS^{0}>0) and nonspontaneous (ΔG^{0}>0). [188]

Nathaniel et al., 2011 study the preparation of organobentonite was conducted using bentonite from Pacitan as the raw material. The modification of bentonite was conducted by microwave irradiation. The adsorption capability of natural and modified-bentonite was tested for the removal of lead metal from aqueous solution. The adsorption experiments were conducted isothermally in static mode at various temperatures. The temperature dependent forms of the Langmuir, Freundlich, Sips and Toth models were used to correlate equilibrium data. It was found that the temperature dependent forms of Sips model can correlate the experimental data better than other models. The pseudo-first-order and pseudo-second-order models were chosen to correlate the experimental kinetic data. The pseudo-second order correlated the experimental data fairly well. [189]

Kapoor, A. and Viraraghavan, 1998 study bentonite was immobilized in a polymer matrix of polysulfone in the form of spherical beads to remove cadmium and copper ions. For this purpose a bead-producing unit was fabricated. The procedure for making beads is described. The beads exhibited a porous surface and surface area of up to 20 m^2/g. 99.4% of beads were retained on a sieve size of 1.18 mm and were finer than 2.36 mm. Batch adsorption data was adequately described by the Freundlich isotherm model. The beads were able to remove more than 99% of copper and cadmium ions in the column mode. These beads can find application as filter media in the removal of toxic heavy metal ions from aqueous solutions. [190]

Okoye and Obi, 2012, reported the Aluminum-Pillared and Unpillared Bentonite Clays were used for removal of heavy metal ion. The transformation of natural clay mineral (Unpillared Bentonite Clays) into pillared clay was achieved using aluminum (0.1M Al $(NO_3)_3.9H_2O$) by adding slowly a basic solution of 0.2M NaOH on the aluminum solution under constant stirring. The detail method is according to Okoye and Obi, 2012. The surface area and porosity was determined using Nitrogen (N2) adsorption-desorption method shows the BET surface area (m^2/g), Micro pore volume (cm^3/g), Micro pore area (m^2/g) and Total pore volume (cm^3/g) values as 29.99, 0.0011, 2.70, and 0.0010 respectively in natural bentonite; and 49.86, 0.0014, 3.30, and 0.013 respectively in Aluminum-Pillared bentonite. The results showed that the maximum sorption of the heavy metals (Cu^{2+}, Cd^{2+}, Hg^{2+}, Pb^{2+}) occurred within 30 minutes at pH of 7-10. Kinetics shows Pseudo-second order model to fit the sorption process. Rate law constant K_2 (g.mg-1.min-1) for unpillared bentonite was found to be 0.18, 0.14, 0.12, and 0.11 in Cu (II), Cd (II), Hg (II), and Pb (II) respectively. While Rate law constant K_2 ($g.mg^{-1}.min^{-1}$) for Aluminum pillared bentonite was found to be 0.41, 0.32, 0.18, and 0.15 in Cu (II), Cd (II), Hg (II), and Pb (II) respectively. The value of R^2 is found to be 0.8 in all the cases. Aluminum pillared bentonite showed higher adsorption rate as well as high capacity for all the metal ions reported in this study, and the order of selectivity are given thus: Cu^{2+}> Cd^{2+}> Hg^{2+} > Pb^{2+}. [191]

This stronger adsorption ability can be attributed to large isomorphous substitution subsequently leading to larger clay interlayer spacing by Al^{7+}. The standard free energy values (ΔG°) obtained from this study are found to be negative and less than -10KJ/mol, indicating that physisorption rather than chemisorption is predominant in the adsorption mechanism. Positive entropy (ΔS°) found in all cases shows a rapid reversible adsorption – desorption phenomena, indicating physisorption of the metal ions on the adsorbent surfaces.

The thermodynamic parameter ΔG°(KJ/mol), ΔH°(J/mol), ΔS°(J/mol.K) values -2.65, +33.10, +13.42 for Cu (II); -2.54, +34.15, +14.12, for Cd (II); -2.46, +36.10, +12.82 for Hg (II) and -2.39, +37.03, +13.27 for Pb (II) in unpillar bentonite at (298 K). While in case of Aluminum pillared bentonite ΔG°(KJ/mol), ΔH°(J/mol), ΔS°(J/mol.K) values are -2.80, +31.80, +12.30 for Cu (II); -2.70, +32.60, +13.10 for Cd (II); -2.55, +29.20, +10.00 for Hg (II) and -2.50, +28.03, +10.77 for Pb (II). The values of ΔH° and Ea were both positive and lie in the range of physisorption. The results clearly suggest that the adsorption of all the metal ions studied in this work is controlled by both surface adsorption and ion exchange phenomena. [191]

Babu et al., 2012, reported the Biosorption performance of bentonite was investigated for the removal of lead from aqueous solutions. Results demonstrated that higher pH (5.0) favored metal ion removal. It is observed that the 88 % removal of Lead by adsorbent bentonite can be better performed at a size of 75µm of bentonite, at a concentration of 100 mg/l, at a pH of 5 and at optimum time of contact 3 hrs. [192]

Rashmi et al.2012, Investigated, the applicability of bentonite and montmorillonite clay mineral intercalated compounds for the removal of total chromium ions from the aqueous solutions in single batch experiments. The clays were modified by treating them with intercalating solution prepared with 100 ml of 0.1 M zirconium oxychloride solution (ZrOCl2 .8H2O) in which 6 N sodium hydroxide (NaOH) aqueous solution was added dropwise. The experiments were conducted under different conditions of pH, initial concentration of metal ions, and dose of clay and interaction time. The maximum binding of Cr^{3+} by zirconium modified montmorillonite (ZrO-M) and ZrO-Bent clay compound (ZrO-M) take place at pH6. It was observed that the ZrO-M shows better capacity of 87.8 % for chromium at pH 6, interaction time of 15 min. compared to that of ZrO-Bent clay compound (ZrO-M) 83.7%. The results of the present investigation show that clay minerals intercalated with zirconium oxychloride are effective adsorbents for the removal of chromium metal from aqueous solutions. [193]

Assameur and Boufatit, 2011 reported the removal of Co^{2+} by acid activated bentonite. The result shows that maximum binding of 98% can be obtained when adsorption take place at pH 5, initial conc. 10ppm, contact time 60min. The binding depend upon pH and shaking time which vary from 5-120min. It shows the adsorption model kinetics best fitted to psedo second order kinetics with R^2=0.99. [194]

Ratanarat et al., 2003, has synthesized the polymeric molecular structure, linear and branch, pol(amidoamine) and their nanocomposites with modified silicate clay. And these nanocomposites for removal of aromatic organic solvents (Toluene and Xylene). It was observed that adsorption ability was much improved from 38% to 98% in toluene and from 22% to 97% in case of xylene by using organically modifies MMT especially 60% MMT of PEO/MMT nanocomposites. The reason for this increase in adsorption is because of higher degree of basal spacing expansion. It was seen that incorporation of PAMAM dendrimer with organoclay has improved the ability to capture organic molecules due to high branched structure with alkyl chain of PAMAM. The adsorption capacity for adsorbing aromatic organic solvents is different in PEO/MMTs and PAMAM/MMT nanocomposites because of polarity and molecular structure of both adsorbent and adsorbate. [195]

Futalan, 2012 reported the competitive removal of copper, nickel and lead from aqueous solution using chitosan immobilized on bentonite (ChB) has been investigated. The presence of Na+ ions was observed to suppress the adsorption capacity in the order of: Ni (II) > Cu (II) > Pb (II). To determine the kinetic rate uptake of Pb(II), Cu(II) and Ni(II) at the solid-liquid interphase of ChB, four kinetic models such as pseudo-first order or Lagergren, pseudo-second order, intraparticle diffusion and film diffusion equation were applied to the experimental data. Kinetics study revealed that adsorption of Cu (II), Ni (II) and Pb (II) in a multi-metal system onto ChB follows the pseudo-second order equation. This signifies that the rate-determining step is chemisorption, which means that covalent bonds are formed through sharing of electrons between ChB and metal ions. The isotherm study shows that Pb(II) and Cu(II) follow Freundlich isotherm while Ni(II) is best described with the Langmuir model. In addition, the values of the Langmuir constant, q , for Cu(II), mL Ni(II) and Pb(II) are 12.6, 6.1 and 15.0 mg g-1, respectively. The preferential adsorption of Pb (II) over Cu (II) and Ni (II) onto ChB is due to properties such as electronegativity, hydrolysis constant and softness value. In the regeneration study, O.1M of HCl, NaCl and NaOH was used. HCl provided the highest desorption capacity but with most material damage to ChB. After three cycles of adsorption-desorption, % desorption using HCl was 79, 70 and 81% for Cu (II), Ni(II) and Pb(II), respectively. [196]

Futalan et al., 2011 study the comparative and competitive adsorption of Cu (II), Ni (II), and Pb (II) from aqueous solution using chitosan immobilized on bentonite (CHB) was investigated. BET surface area (m2/g) 2.12, 84.28, & 33.17 for Chitosan Bentonite & CHB respectively. The Total pore volume (cm3/g) 0.032, 0.085, and 0.048 for Chitosan Bentonite & CHB respectively. The adsorption data of single and binary systems indicated that Cu(II) and Pb(II) best fits Freundlich isotherm while Ni(II) follows Langmuir. In binary systems, a decrease in adsorption capacities and isotherm constants was observed, showing preference of adsorption in the order of Pb (II) > Cu(II) > Ni(II). In single and binary metal systems, Pb(II) has the highest adsorption capacity over Cu(II) and Ni(II). This is influenced by the electronegativity value of a metal ion, where the values are arranged in the order: Pb(II) (2.33)>Cu(II) (1.95)>Ni(II) (1.91). There is more attraction between two atoms when their Electronegativity difference is large. Pb (II) being the most electronegative, results to its preferential adsorption to the hydroxyl (–OH) and amino (–NH2) groups of CHB over Cu (II) and Ni (II) accordingly. The adsorption capacity, qt (mg/g) of Pb, Cu & Ni is 36.01, 23.11 & 17.51 respectively at 25°C. Kinetic studies of single system indicated that the pseudo-second order is the best fit with high correlation coefficients (R^2 > 0.99) and low sum of error values (SE = 0.13–0.46%). Thermodynamic studies illustrated that adsorption of CHB are exothermic and causes a decrease in the entropy. The adsorption of Pb (II) is spontaneous while Ni (II) is non-spontaneous at 25–55 °C. Cu (II) adsorption is only spontaneous at 25 °C. [197]

Calagui et al., 2014, batch adsorption study was utilized in evaluating the potential suitability of chitosan-coated bentonite (CCB) as an adsorbent in the removal of indium ions from aqueous solution. The experimental data were fitted with several isotherm models, where the equilibrium data was best described by Langmuir isotherm. The adsorption capacity is 17.890 mg/g. The correlation coefficient (R^2) is 0.970. The mean energy (E) value was found in the range of 1-8 kJ/mol, indicating that the governing type of adsorption of indium (III) onto CCB is essentially physical. The kinetics was evaluated utilizing the pseudo-first order and pseudo-second order model. The adsorption kinetics of indium (III) best fits the pseudo-second order (R^2 > 0.99), which implies that chemical sorption as the rate-limiting step. Thermodynamic parameters, including Gibbs free energy, enthalpy, and entropy indicated that the indium (III) ions adsorption onto CCB was feasible, spontaneous and endothermic in the temperature range of 278-318 K. The ΔS (KJ/Mol.K)and ΔH(KJ/mol) value 0.0293 and 7.310. [198]

Dalida et al., 2011 examine the batch experiments were executed to investigate the removal of Cu (II) from aqueous solutions using non-cross linked (CCB) and cross linked chitosan-coated (CCB-ECH) bentonite beads. The percentage removal and adsorption capacity of Cu (II) ions were examined as a function of initial concentration and pH. The equilibrium data of CCB agreed well with the Langmuir model while CCB-ECH beads showed a better Fit with the freundlich model. Based on the isotherm study, CCB is a homogenous adsorbent, whereas CCB-ECH is a heterogeneous adsorbent. The adsorption capacities of CCB and CCB-ECH at pH 4 are 12.21 and 9.43 mg/g, respectively. The kinetic data correlated well with the pseudo-second order equation, which implies that chemisorptions is the rate-limiting step. The desorption study was performed using eluent solutions, tap water (pH 7) and HCl solution (pH 1 and pH 3). The best Cu (II) recovery was obtained using HCl solution (pH 1). [199]

Cho et al., 2012 synthesis the composite adsorbent by entrapping cross-linked chitosan and nano-magnetite (NMT) on heulandite (HE) surface to remove Cu(II) and As(V) in aqueous solution. The optimized mass ratio among chitosan, HE, and NMT was determined to be 1:1:2. Kinetics studies indicated the removal of both ions followed pseudo-second order kinetics. The k_2 (g (mg min)-1) is 0.0711 & 0.0269 of Cu (II) and As (V) repectively. The correlation coefficient (R^2) is 0.99. This suggesting specific interaction with surface functional groups was the major route of the removal process. The adsorption capacity of composite gave the maximum equilibrium uptakes of Cu (II) and As (V) of 14.3 and 6.5 mg g1, respectively. The thermodynamic data showed that both adsorption processes were thermodynamically favorable, spontaneous, and endothermic nature. The adsorption capacity for Cu (II) increased continuously with an increase in initial solution pH (3–9), but adsorption of As (V) showed an opposite trend. The overall results demonstrated the potential utility of the composite for Cu (II) and As (V) removal from aqueous solutions. [200]

Ján Vereš1 and Zuzana Orolínová, 2009 first treated the natural bentonite by the sedimentation method. Then the treated bentonite was used to prepare the bentonite/iron oxide composite by the method of co precipitation from the solution of ferric and ferrous salts. The treated and magnetically modified bentonite were prepared and characterized with aim to compare their sorption properties with natural bentonite. The study showed that the bentonite/iron oxide composite and treated bentonite were effective in removing Ni^{2+} ions from aqueous solutions compared to the commercial bentonite from the locality Stará Kremnička. The adsorption capacity on selected sorbents in low concentration range shows following order: composite bentonite/iron oxide composite > treated bentonite > natural bentonite. The adsorption capacity (mg/g) at 5ppm is 12, 6, and 4 in composite bentonite/iron oxide composite, treated bentonite, natural bentonite respectively. [201]

Pandey and Mishra, 2011 reported adsorption of Cr^{6+} by using organic-inorganic composite of chitosan and nanoclay (Cloisite 10A) with combined properties of hydrophilicity of an organic polycation and adsorption capacity of inorganic polyanion. The data fitted satisfactorily to Langmuir isotherms, the R^2 values being 0.998 and 0.999 at 15°C & 35°C respectively indicating unilayer adsorption. Based on Langmuir model, Qmax was calculated to be 357.14 mg/g. The adsorption showed pseudo second order kinetics with a rate constant of 8.0763×10^{-4} g mg($^{-1}$) min($^{-1}$) at 100 ppm Cr(VI) concentration. [202]

D. Saravanan et al., 2013 reported the adsorption of Cr^{6+} ion by using a natural biopolymer chitin/ bentonite from aqueous solution. The authors' studies include batch adsorption equilibrium to study the effect of several parameters such as pH, contact time, adsorbent dose and metal ion concentration. The optimal uptake of free ionic Cr^{6+} occurs at pH 4. The monolayer adsorption capacity qmax (dm^3/g) was found to be 1.810. According to regression coefficient the Freundlich adsorption isotherm (R^2=0.9576) was more fitted for Cr^{6+} than Langmuir (R^2=0.8716). [203] Yang Ming-ping & Fu Yong-jian, 2006 study the organically modify bentonite by using hexadecyltri-methylammonium bromide (CTMAB) for the removal of Cr^{6+} ion from wastewater. The optimum condition for maximum adsorption was suggested as pH 3.0 to 5.0, stirring time about 30 min, and organobentonite dosage 10 g/L. It was seen that 35 mg/L of Cr^{6+} ion in wastewater can remove over 98.0%, using the above optimized conditions.Thus the maximum adsorption efficiency confirm the modified organobentonite as an effective adsorbent for Cr^{6+} removal from wastewater. [204]

Nimrod Bleiman and Yael G. Mishael, 2010 reported adsorption of selenium by using chitosan–montmorillonite composites from water. Chitosan–montmorillonite composites were characterized by XRD, zeta potential, and FTIR measurements. Adsorption isotherms of selenate on the composite, on Al- oxide and on Fe-oxide were in good agreement with the Langmuir model, yielding higher adsorption capacity (qmax) 18.4, 17.2 and 8.2mg/g, respectively. [205]

Karthikeyan and Pius etal.2005, reported removal of Fluoride by montmorillonite clay. Fluoride uptake was found maximum at pH 2, and it decreased with increase in pH. The mate-rial with particle size of 75μm exhibited maximum percentage of fluoride adsorption compared to the other particle sizes. The Langmuir maximum sorption capacity for fluoride was found to be 1.485–1.910 mg/g at different temperatures. Fluoride removal was found to be adversely affected only in the presence of HCO_3^- Adsorption occurred on the surface as well as through intraparti-cle diffusion pattern of the adsorbent material. [206]

Thakre et al., 2010 investigated the chemically modified bentonite clay using magnesium chloride in order to enhance its fluoride removal capacity. It was observed that the MB works effectively over wide range of pH and showed a maximum fluoride removal capacity of $2.26 mgg^{-1}$at an initial fluoride concentration of $5mgL^{-1}$, which is much better than the unmodified bentonite. [207].

From the literatures, though the Removal of Heavy Metal ions by Natural Polymer-silica hybrid materials are many [208-224]. But above reviews done in this chapter cleary shows that recently, bentonite and modified bentonites as adsorbents are gaining wide attention due to their higher potential for inorganic and organic contaminants removal from water.

CONCLUSION

Recently, bentonite and modified bentonites as adsorbents are gaining wide attention due to their higher potential for inorganic and organic contaminants removal from water. In most studies, pseudo-second-order kinetic model was found to fit well with the experimental data. Furthermore, inorganic and organic contaminants sorption by various adsorbents has been found to fit well by Freundlich and/or Langmuir isotherm models.

Results from the recent advances in using bentonites and its modified composites show the flexible nature of the clay and its ecofriendly nature. They are capable of removing organic and inorganic contaminants from drinking water with very high removal ratios of toxic trace metals, nutrients, and organic matter. In most of the cases, they proved to be better or comparable with the existing commercial filter materials, adsorbents, and conventional methods used for decontamination of drinking water. Being natural and their abundance presence makes them a low-cost green, nontoxic adsorbent which can be used for removal of different contaminants from water and making clean and pure drinking water available for developed and developing nations.

Acknowledgments

The author (Dr. Sadanand Pandey) is grateful for the generous financial support received from the National Research Foundation (NRF) and all both the authors thanks library and internet facilities made available by the University of Johannesburg, South Africa.

Abbreviations

Q_{max} : Adsorption capacity
CCB: chitosan-coated bentonite
Al: Aluminum
Si: Silicon
O: Oxygen
FTIR: Fourier transform infrared spectroscopy
DTA-TG: Differential thermal analysis and thermal gravimetric analysis
U: Uranium
mg/L: Milligram per Liter
g/L: Grams per Liter
Cr: Chromium
Fe: Iron
Mn: Manganese
As: Arsenic
mg g^{-1}: Milligrams per gram
μg: Microgram
Cd: Cadmium
Cu: Copper
Hg: Mercury
Pb: Lead
Zn: Zinc
As: Arsenic
NMT: nano-magnetite
CEC: cation exchange capacity
EU: European Union
EPA: Environment Protection Agency
ppb: Parts per billion

mM: Millimolar
WHO: World Health organization
XRD: X-Ray Diffraction
NaOH: Sodium hydroxide
M: Molar
La: Lanthanum
Mg: Magnesium
Zr: Zirconium
Ca: Calcium
HCl: Hydrochloric acid
H_2SO_4: Sulphuric acid
BET Isotherm: Stephen Brunauer, Paul Hugh
Emmett, and Edward Teller theory
KCl: Potassium chloride
Co: Cobalt
Ni: Nickel.

REFERENCES

[1] S.N. Kulshreshtha, "A Global Outlook for Water Resources to the Year 2025". *Water Resources Management*, 12 (3), 167–184(1998).
[2] F.R. Rijsberman, Water scarcity: Fact or fiction? Agricultural Water Management, 80, 5–22(2006).
[3] UN-Water. UN-Water statement on water quality. *www.unwater.org*, (Accessed - 06/10/2011) (2010).
[4] C. Duhigg, The new York times, October 12, (2009).
[5] L. Wang, Animals, News, Water Issues (2013).
[6] D. Gutman, (January 10, 2014). "Freedom Industries execs are longtime colleagues". *The Charleston Gazette*. Retrieved January 10, (2014).
[7] M. Tran,Hungary toxic sludge heading towards the Danube. http://www.guardian.co.uk/world/2010/oct/06/toxic-sludge-hungary-danube? INTCMP =ILCNETTXT3487, (2010).
[8] J. Watts, Chinese environment officials fail to stop oil slick from polluting Yellow river. http://www.guardian.co.uk/environment/2010/jan/04/oil-spill-china-yellow-river (2011).
[9] B. Bryant, Deepwater Horizon and the Gulf oil spill - the key questions answered. http://www.guardian.co.uk/environment/2011/apr/20/deepwater-horizon-key-questions-answered (2011).
[10] K. Kadirvelu, C. Namasivayam, Advances in Environmental Research, 7, 471–478 (2003).
[11] M. Kobya, E. Demirbas, E. Senturk, M. Ince, *Bioresource Technology*, 96, 1518–1521(2005).
[12] T. Halttunen, *Functional Foods Forum, Department of Biochemistry and Food Chemistry, University of Turku*. (2007).

[13] L. Järup, L. Hellström, T. Alfvén, M.D. Carlsson, A. Grubb, B. Persson, C. Pettersson, G. Spång, A. Schütz, C.G. Elinder, *Occupational & Environmental Medicine*, 57, 668-672 (2000).

[14] G. Kazantzis, *Biometals*, 17, 493–498 (2004).

[15] T.S. Nawrot; J.A. Staessen, H.A. Roels, E. Munters, A. Cuypers, T. Richart, A. Ruttens, K. Smeets, H. Clijsters, J. Vangronsveld, *Biometals*, 23, 769–782 (2010).

[16] U.S. EPA 1999. U.S. Environmental Protection Agency. External Review Draft. EPA-453/D-99-002B. Research Triangle Park, NC: Office of Air Quality Planning and Standards. November

[17] M. Ma, X.C. Le, *Clin. Chem.,* 44, 539-550 (1998).

[18] C.J. Chen, H.Y. Chiou, W.I. Huang, S.Y. Chen, Y.M. Hsueh, C.H. Tseng: Systemic non-carcinogenic effects and developmental toxicity of inorganic arsenic. In: Arsenic: Exposure and health effects (Eds: C.O. Abernathy, R.L. Calderon and W.R. Chappell). Chapman and Hall Press, pp. 124-34 (1992).

[19] C.H. Tseng, C.H, *J. Intern. Med. Taiwan,* 10, 224-229 (1999).

[20] N. Oyaro, O. Juddy, E.N.M. Murago, E. Gitonga, *Int. J. Food Agric. Environ.* 5, 119-121(2007).

[21] A.T. Paulino, F.A.S. Minasse, M.R. Guilherme, A.V. Reis, E.C. Muniz, J. Nozaki, *J. Colloid Interface Sci.* 301, 479-487 (2006).

[22] C.E. Borba, R. Guirardello, E.A. Silva, M.T. Veit, C.R.G. Tavares, C.R.G., *Biochem. Eng. J.* 30,184-191(2006).

[23] H.C. Yanga, Y.J. Choa, H.C. Eunb, E.H. Kima, *Chemical Engineering Science,* 62 , 5137 – 5143 (2007).

[24] D. Suteu, C. Zaharia, D. Bilba, A. Muresan, R. Muresan, A. Popescu, *Industria Textila*, 60, 254-263 (2009).

[25] C. Zaharia, D. Suteu, A. Muresan, R. Muresan, A. Popescu, *Environmental Engineering and Management Journal,* 8, 1359-1369 (2009).

[26] T. Tanaka, O. Takahashi, S. Oishi, A. Ogata, *Reproductive Toxicology,* 26, 156–163 (2008).

[27] Y.T. Wang, F.L. Zhao, K.A. Li, S.T. Tong, *Analytica Chimica Acta*, 396,75–81(1999).

[28] Y.N. Ni, X.Z. Zhong, *Chinese Chemical Letters,* 18(5), 569-572(2007).

[29] C.Z. Huang, Y.F. Li, P. Feng, *Talanta,* 55, 321–328 (2001).

[30] H. Heli, S.Z. Bathaie, M.F. Mousavi, *Electrochemistry Communications*, 6, 1114–1118 (2004).

[31] G. Crini, *Progress in Polymer Science,* 30, 38–70 (2005).

[32] K.L. Salipira, B.B. Mamba, R.W. Krause. T.J. Malefetse, S.H. Durbach, *Environmental Chemistry Letters,* 5, 13–17 (2007).

[33] H. Zhang, Z. Jin, L. Han C. Qin, *Transactions of Nonferrous Metals Society of China, 16,* 345-349 (2006).

[34] V. Tanboonchuy, J.C Hsu, N. Grisdanurak, C.H. Liao, *World Academy of Science, Engineering and Technology* , 65 , 263 – 265 (2010).

[35] A. Fujishima, T.N. Rao, D.A. Tryk, Photochemistry Reviews, 1, 1–21(2000).

[36] W.Y. Ahn, S.A. Sheeley, T. Rajh, D.M. Cropek, *Applied Catalysis B: Environmental,* 74, 103–110 (2007).

[37] A.C. Baumgardner, M.Sc. Dissertation, Arizona State University, (2005).

[38] K. Hristovski, P. Westerhoff, T. Möller, P. Sylvester, W. Condit, H. Mash, *J. Hazardous Materials.* 152 (1), 397 – 406 (2008).
[39] D.P.E. Dickson, S.A. Walton, NanoStructured Materials, 9, 595-598 (1997).
[40] D.R. Paul, L.M. Robeson, *Polymer* 49, 3187-3204 (2008).
[41] L.A. Utracki, M.R. Kamal, *The Arabian Journal for Science and Engineering* 27, 43-67 (2002).
[42] M. Rosorff, Polymer-Clay Nanocomposites. In *Nano Surface Chemistry*, Marcel Dekker Inc: New York and Basel, 653-673 (2002).
[43] M. Okamoto, *Rapra Review Reports* 14, 1-40 (2003).
[44] M. Alexandre, P. Dubois, *Materials Science and Engineering: A* 28, 1-63(2000).
[45] S. Ray, M. Okamoto, *Progress in Polymer Science* 28, 1539-1641(2003).
[46] M.W. Noh, D.C. Lee, *Polymer Bulletin* 42, 619-626 (1999).
[47] S. Yariv, H. Cross, Organo-Clay Complexes and Interactions. In Marcel Dekker, Inc: New York and Basel, 1-101(2002).
[48] M. Huskic, I. Brnardic, M. Zigon, M. Ivankovic, *Journal of Non-Crystalline Solids* 354, 3326-3331(2008)
[49] S.P. Parker ed. McGraw-Hill encyclopedia of the geological sciences, 2nd ed. New York, McGraw-Hill, pp 32–33, 69–72, 400–401(1988).
[50] R.E. Grim, Clay mineralogy, 2nd ed. New York, McGraw-Hill, 596 (1968)
[51] S.H. Patterson, H.H. Murray, Clays. In: Lefond SI ed. Industrial minerals and rocks,5th ed. New York, American Institute of Mining, Metallurgical, and Petroleum Engineers, 585–651(1983).
[52] J.W. Hosterman, S.H. Patterson. Bentonite and Fuller's earth resources of the United States. U.S. Geological Survey Professional Paper 1522. United States Government Printing Office, Washington D.C., USA (1992).
[53] R.E. Bates, J.A. Jackson ed., Glossary of geology, 3rd ed. Alexandria, Virginia, American Geological Institute, pp 65, 123, 262, 432(1987).
[54] S.W. Bailey chairman, Summary of recommendations of AIPEA [Association Internationale Pour l'Étude des Argiles] nomenclature committee on clay. *Am. Mineral*, 65: 1–7 (1980).
[55] M. Rieder, G. Cavazzini, Y.S. D'yakonov , V.A. Frank-Kamenetskii , G. Gottardi, S. Guggenheim , P.V. Koval, G. Müller , A.M.R. Neiva, E.W. Radoslovich, J-L. Robert , F.P. Sassi, H. Takeda, Z. Weiss, D.R.Wones, *Clays Clay Miner,* 46: 586–595 (1998).
[56] G. Beyer, *Plast. Addit. Compound*, 4(10), 22–7(2002).
[57] P. Liu, L. Zhang, *Separation and Purification Technology*, 58, 32-39 (2007).
[58] A. Meunier, Clays. Heidelberg: Springer-Verlag Berlin (2005).
[59] L. Wang, A. Wang, *Journal of Hazardous materials*, 160, 173-180 (2008).
[60] B.A. Kennedy, *Surfacing Mining*, 2nd Edition, Published by Society for Mining, Metallurgy, and Exploration (1990).
[61] M. Ahmaruzzaman, *Advances in Colloid and Interface Science*, 143, 48–67 (2008).
[62] B.H. Jeong, E.M.V. Hoek, Y. Yan, A. Subramani, X. Huang, G. Hurwitz, A.K. Ghosh, J. Jawor, *Journal of Membrane Science*, 294, 1–7 (2007).
[63] R.A. Khaydarov, R.R. Khaydarov, O. Gapurova, *Water Research*, 44, 1927 – 1933 (2010).
[64] E. Montarges, L.J. Michot, F. Lhote, T. Fabien, F. Villieras, *Clays and Clay Minerals,* 43 (4), 417-426 (1995).

[65] Y. Ku, I.L. Jung, *Water Res.* 35, 135-142 (2001).
[66] K.A. Baltpurvins, R.C. Burns, G.A. Lawrance, A.D. Stuart, *Water Res.* 31, 973-980 (1997).
[67] S.A. Mirbagheri, S.N. Hosseini, *Desalination* 171, 85-93 (2005).
[68] L. Charerntanyarak, *Water. Sci. Technol.* 39, 135-138 (1999).
[69] D. Bhattacharyya, A. B. Jumawan Jr., R. B. Grieves' *Separation Science and Technology* ,14, (5),441-452 (1979).
[70] Özverdi, M. Erdem, *J. Hazard. Mater.* 137, 626-632 (2006).
[71] a) M.M. Matlock, K.R. Henke, D.A. Atwood, J. Hazard. Mater. 92, 129-142 (2002a). b)M.M. Matlock, B.S. Howerton, M.A.V. Aelstyn, F.L. Nordstrom, D.A. Atwood, *Environ. Sci. Technol.* 36, 1636-1639 (2002b).c)M.M. Matlock, B.S. Howerton, D.A. Atwood, *Water Res.* 36, 4757-4764 (2002c).
[72] F.L. Fu, R.M. Chen, Y. Xiong, *Sep. Purif. Technol.* 52, 388393(2006).
[73] F.L. Fu, H.Y. Zeng, Q.H. Cai, R.L. Qiu, J. Yu, Y. Xiong, *Chemosphere* 69, 1783-1789 (2007).
[74] Y. Xu, T. Xu, *IEEE,* 2789-2793 (2008).
[75] Y. Xu, Z. Xie , L. Xue. *J. Environ. Sci. (China)* 23(5):778-83 (2011).
[76] S.Y. Kang, J.U. Lee, S.H. Moon, K.W. Kim, *Chemosphere* 56, 141-147 (2004).
[77] S.A. Abo-Farha, A.Y. Abdel-Aal, I.A. Ashourb, S.E. Garamon,. *J. Hazard. Mater.* 169, 190-194 (2009).
[78] T. Motsi, N.A. Rowson, M.J.H. Simmons, *Int. J. Miner. Process* 92, 42-48 (2009).
[79] I.C. Ostroski, M.A.S.D. Barros, E.A. Silvab, J.H. Dantas, P.A. Arroyo, O.C.M. Lima, *J. Hazard. Mater.* 161, 1404-1412 (2009).
[80] S.R. Taffarel, J. Rubio, *Miner. Eng.* 22, 336-343(2009).
[81] M.K. Doula, A. Dimirkou, *J. Hazard. Mater.* 151, 738-745 (2008).
[82] M.K. Doula, *Water Res.* 43, 3659-3672 (2009).
[83] A.G. Bayer Removing traces treating industrial waste water, Process water and groundwater with Lewatit ion exchange resins. Bayer AG, Leverkusen (2000)
[84] K. Lee, J. Hong, *AIChE J.* 41, 2653–2657 (1995).
[85] C. Colella, Use of italian chabazite and phillipsite for the removal of heavy metals from wastewaters. In: Ming, D.W., Mumpton F.A. (Eds.), Natural Zeolites 93, Int. Comm, Natural Zeolites Brackport, New York, p. 363 (1995).
[86] M. Pansinini, C. Colella, D. Caputo, M. de'Gennaro, A. Langella, *Micropor. Mater.* 5, 357–364 (1996).
[87] E. Torracca, P. Galli, M. Pansini, C. Colella, *Micropor. Mesopor. Mat.* 20, 119–127 (1998).
[88] W.C. Gardiner, F. Munoz,. *Eng.* 78, 57–59 (1971).
[89] G.J. De Jong, G.J.N. Rekers, *J. Chromatogr.* 102, 443–450 (1974).
[90] M. Ocenasek, Z. Matejka, V. Kriz, Sb. Vys. Sk. *Chem.—Technol. Praze Technol. Paliv.* D 41, 43–64 (1980).
[91] B.A. Bolto, L. Pawłowski, Wastewater Treatment by Ion-Exchange. E. and F.N. Spoon Ltd., London (1987).
[92] A. Warshawsky, Chelating ion exchangers. In: Streat, M., Naden, D. (Eds.), Ion Exchange and Sorption Processes in Hydrometallurgy. John Wiley and Sons, Chichester–New York–Brisbane–Toronto–Singapore (1987).
[93] K. Dorfner, K. (Ed.), Ion-Exchangers. Walter de Gruyter, Berlin-New York (1991).

[94] D. Kauspediene, J. Snukiskis, A. Gefeniene,. *J. Radioanal. Nucl. Chem.* 229, 129–132 (1998).

[95] K. Halle, K. Fischwasser, B. Fenk, *Technol. Umweltschutz* 25, 120– 132 (1982).

[96] I. Untea, E. Tudorache, C. Florea, *Revista de Chimie* 51, 123–126 (2000).

[97] S.E. Kurashvili, A.N. Barash, G.D. Litovchenko, *Fibre Chem.* 29, 149–152 (1997).

[98] E. Fourest, J.C. Roux, *Applied Microbiology and Biotechnology* , 37, 399-403 (1992).

[99] R. Apiratikul, P. Pavasant, *Chem. Eng. J.* 144, 245-258 (2008).

[100] D. Kratochvil, B. Volesky, *Trends Biotechnol* 16, 291–300 (1998).

[101] T. Aman, A.A. Kazi, M.U. Sabri, Q. Bano, *Colloid Surf.* 63, 116-121(2008).

[102] F. Kaczala, M. Marques, W. Hogland, *Bioresour. Technol.* 100, 235-243(2009).

[103] A. Saeed, M. Iqbal, M.W. Akhtar, J. Hazard. Mater. 117, 65-73(2005).

[104] P.H. Jai, J.S. Wook, Y.J. Kyu, K.B. Gil, L.S. Mok,. *J. Environ. Sci.* 19, 1436-1441(2007).

[105] O.S. Amudaa, F.E. Adelowoa, M.O. Ologunde, *Colloid Surf.* 68, 184-192 (2009).

[106] L. Oliveira, A.S. Franca, T.M. Alves, S.D.F. Rocha, *J. Hazard. Mater.* 155, 507-512 (2008).

[107] Y.N. Mata, M.L. Blázquez, A. Ballester, F. González, J.A. Muñoz, *Chem. Eng. J.* 150, 289-301(2009).

[108] S. Schiewer, S.B. Patil, *J. Hazard. Mater.* 157, 8-17 (2008).

[109] L.C. Ajjabi, L. Chouba, *J. Environ. Manage.* 90, 3485-3489 (2009).

[110] P. Pavasant, R. Apiratikul, V. Sungkhum, P. Suthiparinyanont, S. Wattanachira, T.F. Marhaba, *Bioresour. Technol.* 97, 2321-2329 (2006).

[111] A. El-Sikaily, A.E. Nemr, A. Khaled, O. Abdelwehab, *J. Hazard. Mater.* 148, 216-228 (2007).

[112] L.P. Deng, Y.Y. Su, H. Su, X.T. Wang, X.B. Zhu,. *J. Hazard. Mater.* 143, 220-225 (2007).

[113] J.H. Pan, R.X. Liu, H.X. Tang, *J. Environ. Sci.* 19, 403-408 (2007).

[114] M. Souiri, I. Gammoudi, H.B. Ouada, L. Mora, T. Jouenne, N. Jaffrezic-Renault, C. Dejous, A. Othmane, A.C. Duncan, *Proced. Chem.* 1, 1027-1030 (2009).

[115] C. Quintelas, Z. Rocha, B. Silva, B. Fonseca, H. Figueiredo, T. Tavares, *Chem. Eng. J.* 152, 110-115 (2009).

[116] R.M. Gabr, S.H.A. Hassan, A.A.M. Shoreit, *Int. Biodeterior. Biodegradation* 62, 195-203 (2008).

[117] H. Salehizadeh , S.A. Shojaosadati, *Water Res.* 37,4231–5 (2003).

[118] S.B. Choi , Y.S. Yun, *Biotechnol. Lett.* 26,331–6 (2004).

[119] J.S. Chang, R. Law , C.C. Chang,*Water Research.* 31,1651–8(1997).

[120] G.Uslu , M. Tanyol, *J. Hazard. Mater.* 135,87–93(2006).

[121] A. Selatnia, A. Boukazoula, N. Kechid, M.Z. Bakhti, A.Chergui, Y. Kerchich, *Biochem. Eng. J.* 19,127–35(2004).

[122] Incharoensakdi A, Kitjaharn P. Zinc biosorption from aqueous solution by a halotolerant cyanobacterium Aphanothece halophytica. *Curr. Microbiol.* 2002;45:261–4.

[123] X.C. Chen,Y.P Wang, Q. Lin, J.Y. Shi,W.X. Wu, Y.X. Chen, *Colloid Surf B-Biointerfaces* 4 6, 101–7 (2005).

[124] N. Mameri , N. Boudries , L. Addour , D. Belhocine , H. Lounici , H. Grib et al. *Water Res* 33, 1347–54 (1999).

[125] H.L. Liu HL, B.Y. Chen BY, Y.W. Lan YW, Y.C. Cheng, *Chem. Eng. J.* 97,195–201(2004).
[126] A. Nakajima, M. Yasuda, H. Yokoyama , H. Ohya-Nishiguchi , H. Kamada, *World J. Microbiol. Biotechnol.* 17, 343–7 (2001).
[127] W.B. Lu, J.J. Shi, C.H. Wang , J.S. Chang, *J. Hazard. Mater.* 134,80–6 (2006).
[128] I. Savvaidis I, M.N. Hughes MN, R.K. Poole,*World J. Microbiol. Biotechnol.* 19,117–21(2003).
[129] G. Uslu , M. Tanyol, *J. Hazard. Mater.* 135,87–93 (2006).
[130] A. Ozturk A, T. Artan T, A. Ayar, Colloid Surf B-Biointerfaces 34, 105–11(2004).
[131] M.X. Loukidou, T.D. Karapantsios , A.I Zouboulis ,K.A. Matis, *J. Chem. Technol. Biotechnol.* 79,711–19 (2004).
[132] M. Ziagova , G. Dimitriadis , D. Aslanidou , X. Papaioannou , E.L. Tzannetaki , M. Liakopoulou- Kyriakides , *Bioresour. Technol.* 98,2859–2865 (2007).
[133] A. Ozturk, *J. Hazard. Mater.* 147, 518–523 (2007).
[134] I. de Vargas , L.E. Macaskie , E. Guibal, *Chem. Technol. Biotechnol.* 79:4 9–56 (2004).
[135] A. Nakajima A, T. Tsuruta, *J. Radioanal. Nucl. Chem.* 260, 13–18 (2004).
[136] R. Say , N. Yimaz , A. Denizli, Sci Technol 21, 643–50 (2003).
[137] H. Niu , X.S. Xu , J.H. Wang , B. Volesky, *Biotechnol. Bioeng.* 42,785–787 (1993).
[138] S.B. Deng , Y.P Ting, Langmuir 21,5940–8(2005).
[139] A. Kapoor , T. Viraraghavan, *Bioresour. Technol.* 53,195–206(1995).
[140] a)A.Y. Dursun, Fresenius Environ Bull 12,1315–22 (2003a).b)A.Y Dursun AY, G. Uslu G, O. Tepe O, Y. Cuci Y, H.I. Ekiz, *Biochem. Eng. J.* 15,87–92(2003b).
[141] N.N. Lyalikova-Medvedeva, T.V. Khijniak, Biosorption of long-lived radionuclides. Biohydrometallurgy and the environment toward the mining of the 21st century, Pt B 1999. Process MetallurgyAmsterdam: *Elsevier Science Publ.* 327–34(1999).
[142] J. Landaburu-Aguirre, V. García, E. Pongrácz, R.L. Keiski, *Desalination* 240, 262-269(2009).
[143] E. Sampera, M. Rodrígueza, M.A. De la Rubia, D. Prats, *Sep. Purif. Technol.* 65, 337-342 (2009).
[144] J. Labanda, M.S. Khaidar, J. Llorens, *Desalination* 249, 577-581(2009).
[145] M.K. Aroua, F.M. Zuki, N.M. Sulaiman, *J. Hazard. Mater.* 147,752-758 (2007).
[146] R. Molinari, T. Poerio, P. Argurio, Chemosphere 70, 341-348 (2008).
[147] K. Trivunac, S. Stevanovic, *Chemosphere* 64, 486-491(2006)
[148] H.J. Kim, K. Baek, B.K. Kim, J.W. Yang, *J. Hazard. Mater.* 122, 31-36(2005).
[149] M. Mohsen-Nia, P. Montazeri, H. Modarress, Desalination 217, 276-281.(2007)
[150] R. D. Letterman (ed.) "Water Quality and Treatment." 5th Ed. (New York: American Water Works Association and McGraw-Hill.) ISBN 0-07-001659-3 (1999).
[151] M. Muthukrishnan, B.K. Guha, *Desalination* 219, 171-178(2008).
[152] E. Cséfalvay, V. Pauer, P. Mizsey, *Desalination* 240, 132-142 (2009).
[153] H. Koseoglu, M. Kitis, *Miner. Eng.* 22, 440-444(2009).
[154] C.M. Nguyen, S. Bang, J. Cho, K.W. Kim, Desalination 245, 82-94(2009).
[155] P. Erikson, *Environ. Prog.* 7, 58-61(1988).
[156] M. Sadrzadeha, T. Mohammadi, J. Ivakpour, N. Kasiri, *Chem. Eng. Process.* 48, 1371-1381(2009).
[157] J. Lambert, M. Avila-Rodriguez, G. Durand, M. Rakib, *J. Membr. Sci.* 280, 219-225(2006).

[158] T. Mohammadi, A. Razmi, SM. Adrzadeh, *Desalination* 167, 379-385(2004).
[159] Heidmann, W. Calmano, *J. Hazard. Mater.* 152, 934-941(2008).
[160] T. Ölmez, *J.Hazard.Mater.*162, 1371-1378(2009).
[161] W.H.Glaze, J-W. Kang, D.H. Chapin, *Ozone: Science & Engineering: The Journal of the International Ozone Association* 9 (4), 335–352 (1987).
[162] A. Matilainen, M. Sillanpää , *Chemosphere* 80(4), 351-65(2010).
[163] R.M. Hozalski, E.J. Bouwer, S. Goel, *Water Science and Technology,* 40(9) 157–163(1999).
[164] S. Pandey S. B. Mishra, S. B. Journal of Sol-Gel Science and Technology, 59 (1) 73-94 (2011).
[165] G. Annadurai, R.S. Juang, D.J. Lee, *Journal of Hazardous Materials*, B92, 263–274(2002).
[166] Z. Shipeng, C.Jinyao, L. Huilin, *Polymer Bulletin* 63, 245-257(2009)
[167] C.Y. Young; L.S Hyoung; R.S, *Polymer Bulletin* 63, 47-55(2009) .
[168] G. Zehetmeyer, J.M. Scheibel, R.M.D. Soares, *Polymer Bulletin* 70, 2181-2191(2013)
[169] J. Pascual, E. Fages, O. Fenollar, *Polymer Bulletin* 62,367-380 , (2009).
[170] Y. Xu, Q. Chen, W. Bai, *Polymer Bulletin* 68,983-992(2012) .
[171] X. Li, Z. Zhan, G. Peng, *Polymer Bulletin* 67, 719-727(2011) .
[172] F.A. Fahimeh, O. Ali, M. Abdolreza, *Polymer Bulletin* 1-20 (2014)
[173] J.Q. Jiang , Z. Zeng, *Chemosphere* 53, 53-62 (2003)
[174] K. G. Bhattacharyya, S. Sen Gupta, *Advances in Colloid and Interface Science*, 140, 114–131 (2008).
[175] M. J. Angove, B. B. Johnson, J. D.Wells, *Journal of Colloid and Interface Science*, 204, 93–103(1998).
[176] D.J.L. Guerra, I. Mello, R. Resende, R. Silva , *Water Resources and Industry* 4,32–50(2013).
[177] S. Triantafyllou, E. christodoulou, P. neou-syngouna, *Clays and Clay Minerals,* 47(5), 567-572 (1999).
[178] H. A. Talaat, N.M. E Defrawy, A.G. Abulnour, H.A. Hani, *2nd International Conference on Environmental Science and Technology*, *IPCBEE* .6 *(2011) IACSIT Press, Singapore*
[179] F. Mohammed-Azizi, S. Dib, M. Boufatit, *Desalination and Water Treatment,* 51, 4447–4458(2013).
[180] Kaya, A.H. Oren , *Journal of Hazardous Materials* B125, 183–189(2005).
[181] J.A. Hefne, W.K. Mekhemer, N.M. Alandis, O.A. Aldayel, T. Alajyan, JKAU: Sci., 22(1) 155-176 (2010).
[182] F. Ayari, E. Srasra, M. Trabelsi-Ayadi, *Asian Journal of Chemistry,* 19(5), 3325-3339 (2007).
[183] B. I. Olu-Owolabi , D. B. Popoola , E. I. Unuabonah, *Water Air Soil Pollut* 211(1), 459-474 (2010)
[184] A. Bourlival,K. Michailidis, C. Sikalidis, A. Filippidis, M. Betsiou, *Fresenius environmental bulletin*, 21, 8c.(2012).
[185] W. Bekele, G. Faye, N. Fernandez, International journal of research in pharmacy and chemistry, 4(1), 192-201 (2014).
[186] S.S. Al-Shahrani, *Alexandria Engineering Journal,* 53, 205–211(2014).
[187] H. Bedelean, A. Măicăneanu, . Burcă, M. Stanca, *Clay Minerals,* 44,. 487-495(2010).

[188] G.G. Sandy, V. Maramis, A. Kurniawan, A. Ayucitra, , S. Ismadji , *Frontiers of Chemical Science and Engineering* , 6(1), 58-66(2012).
[189] E. Nathaniel, A. Kurniawan, F.E. Soeteredjo, S. Ismadji, *Desalination and Water Treatment*, 36, 280–288(2011).
[190] A. Kapoor, T. Viraraghavan, *J. Environ. Eng.*, 124(10), 1020–1024 (1998).
[191] I.P. Okoye, C. Obi, *International Archive of Applied Sciences and Technology* 3(2) 58 – 67(2012).
[192] D.J. Babu, G.Kalyani, H.J. Rao, Y.P. Kumar, P.King, *International Journal of Scientific & Engineering Research,* 3(8) 2012.
[193] O. Rashmi , D.S. Ramteke P.U. Meshram, *Journal of Research in Chemistry and Environment* 2(2) ,83-87(2012).
[194] H. Assameur, M. Boufatit, Proceeding of global warming, 11-14 july, Lisbon, Portugal (2011).
[195] K. Ratanarat, M. Nithitanakul, D.C. Martin, R. Magaraphan, *Rev.Adv. Mater.sci.*5, 187-192(2003).
[196] C.M. Futalan, W.-C. Tsai, S.-S. Lin, K J Hisen , M.L. Dalida, M.-W. Wan, *Sustain. Environ. Res.,* 22(6), 345-355 (2012).
[197] C.M. Futalan[a], C-C. Kan[b], M. L. Dalida[c], K-J. Hsien, C. Pascua[d], M-W. Wan, Carbohydrate Polymers , 83(2), 528–536 (2011).
[198] M.J. C. Calagui, D. B. Senoro, C.-C. Kan, J.W.L. Salvacion, C.M. Futalan, M.-W. Wan, doi- dx.doi.org/doi:10.1016/j.jhazmat.2014.04.043(2014).
[199] M. L. P. Dalida, A.F. V. Mariano, C. M. Futalan, C.-C. Kan, W.-C.Tsai, M.-W. Wan, *Desalination*, 275(1–3), 154-159(2011).
[200] D.-W. Cho, B.-H. Jeon, C.-M. Chon, Y. Kim, F. W. Schwartz ,E.-S. Lee, H. Song, *Chemical Engineering Journal*, 200–202, 654-662(2012).
[201] J. Vereš, Z. Orolínová, *Acta Montanistica Slovaca Ročník* 14, 152-155(2009).
[202] S. Pandey, S. B. Mishra, *Journal of Colloid and Interface Science*, 361(2), 509–520 (2011).
[203] D. Saravanan ,T. Gomathi, , P.N. Sudha, *International Journal of Biological Macromolecules* 53, 67– 71(2013).
[204] Y. Ming-ping, F. Yong-jian, Materials Protection (2006)-02.
[205] N. Bleiman, Y. G. Mishael, *Journal of hazardous material* , 183(1-3), 590-595(2010).
[206] G. Karthikeyan, A. Pius, G. Alagumuthu, *J. Chem. Technol.* 12, 263–272 (2005).
[207] D. Thakre, S. Rayalu, R. Kawade, S. Meshram, J. Subrt, N. Labhsetwar, *Journal of Hazardous Materials,* 180, 122–130 (2010).
[208] V. Singh, S. Pandey, S.KSingh, R. Sanghi, *Sep. Purif. Technol.* 67, 251–261(2009).
[209] V. Singh, S. Pandey, S.KSingh, R. Sanghi, *J. Sol-Gel Sci. Technol.* 47, 58–67 (2008).
[210] V. Singh, A.Tiwari, S. Pandey, S. K Singh, R. Sanghi, *J. Appl. Polym. Sci.* 104, 536–544 (2007).
[211] S. Pandey, G.K. Goswami, K.K. Nanda, *Scientific Reports* (3), doi: 10.1038/srep02082) (2013).
[212] S. Pandey, G.K. Goswami, K.K. Nanda, *Carbohydrate Polymers* 94, 229-234 (2013).
[213] S. Pandey, S.B. Mishra, *Carbohydrate Polymer.* 92, 2201– 2205 (2013).
[214] S. Pandey, G.K. Goswami, K.K. Nanda, *International Journal Of biological Macromolecules*, 51, 583-589 (2012).
[215] S. Pandey, S. B. Mishra, *Carbohydrate Polymer.* 90 (1), 370-379, (2012).

[216] S. Pandey, S.B. Mishra, *International Journal of Biological macromolecules.* 49 (4), 527–535 (2011).

[217] V. Singh, A. Tiwari, S. Pandey, S. K. Singh, *xPress polymer letters* 1(1), 51-58 (2007).

[218] V. Singh, A. Tiwari, S. Pandey, S.K. Singh, *Starch/Starke*, 58, 536-543 (2006).

[219] S. Pandey, S. B. Mishra, Amorphous mixed oxides: A New and Highly Versatile Class of Materials, *Intelligent nanomaterial.* 149, 2012 ISBN: 978-0-470-93879-9 Scrivener-Wiley publisher (2012).

[220] S. Pandey, S. B.Mishra, Bioceramics: Silica based organic-inorganic hybrid materials for medical applications, "*Nanomedicine for Drug Delivery and Therapeutics*". ISBN: 9781118414095, Chapter-5, pg-136-158, Scrivener-Wiley publisher (2013).

[221] S. Pandey, S.B Mishra, Chemical Nanosensors for Monitoring Environmental Pollution. *Application of Nanotechnology in Water Research*, 309-332 (2014).

[222] V. Singh, S. Pandey, S.K Singh, R. Sanghi, *Journal of Fudan University (Natural Science)* 5, 155 (2007).

[223] V.Singh, S.K. Singh, S. Pandey, R. Sanghi, *International Journal of Biological Macromolecules* 49 233–240 (2011).

[224] V. Singh, S.K. Singh, S.Pandey, P. Kumar, *Journal of Non-Crystalline Solids* 357, 194–201(2011).

In: Bentonite
Editor: Ajay Kumar Mishra
ISBN: 978-1-63482-142-1

Chapter 5

APPLICATIONS OF BENTONITE AND ITS ROLE AS A BASE OF ENGINEERING CHEMISTRY

Tanvir Arfin
Environmental Materials Division, CSIR-National Environmental Engineering Research Institute (CSIR-NEERI), Nehru Marg, Nagpur, India

ABSTRACT

Bentonite is a layer of natural silicate clay mineral corresponding to a numerous of underlying properties such as large specific surface area, great absorption and interlayer exchangeable cationic properties etc. Betonite is also a hybrid photocatalyst which can increase the effect of adsorption of organic matter, and it is also responsible for increasing the light absorption mechanism, especially in and around the visible region. It has the capability to break the continuous ion conducting phase of the alkaline solid polymer electrolyte (ASPE). It has both positive as well as negative effects on the ionic conductivity of ASPE and also found to decrease the ASPE conductivity. It even has good stability in alkaline medium. The clay of bentonite is generally in a form of montmorillonite and shows an intrinsic cation exchange property, which is not dependent on pH. It is originating from Wyoming or Texas which tends to have either sodium or calcium ion as the principle exchangeable cations. The hydraulic property of the compact bentonite is used as an engineered barrier material which is the basic property for designing the disposal system.

INTRODUCTION

Bentonite in a lucid way is defined as a sedimentary rock which consists of a large proportion of comprehensive clay minerals having a three-layer structure called smectites, namely montmorillonite, beidellite, and non-tronite. The different types of minerals such as quartz, feldspars, zeolites which comprises Bentonite are framed in the formula as $Na_x(H_2O)_4(Al_{2-x}Mg_{0.83})Si_4O_{10}(OH)_2$; it has an underlying applauding property of higher water absorption ability of up to 10 times its weight than other compounds and it also acts as a good stabilizer in alkaline medium. On the basis of above mentioned properties, bentonite is considered as the probable inorganic content required for the preparation of alkaline polymer composite along with higher conductivity [1]. Many scientists gave their view for bentonite and some of them are stated below. Bentonite is regarded as a silicate clay mineral which is naturally layed, has prospective properties such as large specific surface area, interlayer exchangeable cation and great absorption characteristic and also as a carrier, it has an ability of combining with nano-ZnO to form an efficient composite photocatalyst [2,3] for organic treatment in aqueous state. Bentonite in composite photocatalyst increases the effect of adsorption of organic matter and it also increases light absorption, most probably in the visible light [4].

Bentonite as a montmorillonite clay displaying a property of intrinsic cation exchange considerably independent of pH [5-7]. The intrinsic cation is found to be dependent upon the source of the clay. Bentonites which have been derived from Wyoming or Texas contribute to have Na^+ and Ca^{2+} respectively as the most promising exchangeable cations [5-7]. It is also noticed that such native cation is replaced by the substance to be experimented in the laboratory ion-exchange procedure by different sources such as by passing a suspension through a cation exchanger colume or by washing the bentonite employing the aqueous solution of strongly wished cation repeatedly for several times. By employing the second method of clay processing [8], and advance range of eco- friendly bentonite pigments, which promisingly shows the conventional salt-spray tests on complete formulated organic coated galvanized steel [9,10].

It was also evidently shown that the model had an organic coating randomly based on bentonite pigments which contain trivalent rare earth [11,12] and alkaline earth cations and apparently found to have an intrinsic ability for obstructing the corrosion-driven cathodic disbandment on zinc and iron surfaces as well denoting its significance. Such cations are non-toxic and it generally explores the corrosion through the deposition of films that are highly capable of stifling cathodic oxygen reduction [13-18]. Due to its high swelling capacity, bentonite is chosen as a buffer and backfill material for an engineered barrier. For the heating effect of the nuclear waste, the engineered clay barrier and adjacent host rock are responsible and it is likewise connected with hydraulic and mechanical phenomena generally interacting in a complex pattern. The compact bentonite is originally unsaturated but later on, action to hydration from the surrounding rock it triggers and couple with thermo-hydro-mechanical (THM) phenomena. There should be a good relation and understanding between the processes for attaining a secure and long lasting design and structure which generally take place in the surrounding fields and even during the period of evolution.

Till date, various types of bentonites have been distinguished as a specific material for engineer's barrier in a high level waste (HLW) repository. In Swden [19], Germany [20], and France [21], the MX-80 bentonite was recognized as disposal concepts such as backing and sealing material. The FEBEX bentonite, which was selected by ENRESA, the Spanish (the Spanish Agency for Radioactive Waste Management was found to be extracted from the Cortiji de Archidona deposit generally at the zone of serrata de Níjar in Almerí, Spain. It was extensively appropriate for the backfilling and for the sealing of HLW repositories [22]. Along with the above two bentontes, the Kunigel-V1 and FoCa bentonite was also initiated to be probable buffer material in Japan and in France [23-24].

SWELLING PROPERTY

Bentonite based materials have two attributes, namely swelling pressure and swelling potential. A peculiar swelling pressure, such as isotropic one is accumulated for the designing of sealing plugs or engineered barrier and even in the evaluation of swelling behaviour. In the laboratory measurements, the one and only one swelling pressure is concentrated which is the axial one.

Different physicist have performed studies on various levels to describe the swelling properties of Bentonite –based material [25-31] which was recommended to be highly dependent on the initial state including the dry density and water content [28,32].

In the testing ground, various methods were employed to find the swelling pressure. The methods include constant volume, swell-load, zero swell and adjusted constant volume methods [33] that have high capability to be measured in the axial direction. Researchers have even attempted to get the radial swelling pressure [34-36]; for example, Lee et al. [36] have measured radial and axial, both the swelling pressures at constant volume cell with pressure sensors that was ascended in the two directions where the difference between the radial and axial swelling pressures was noticed tremendously with the largest difference at the higher dry densities. This was described by the anisotropic microstructure of compact bentonite in a motivated sequence.

When it has been installed in the field, in some geological host formations which is relatively rich in salts such as salt formation, clay stone, granite formations, the site water of certain salinity gain it possessed an ability to interact with bentonite [37]. Due to which the hydro-mechanical behaviour of the bentonite-based materials can be hindered and the barrier performance may also change leading to the alteration in water composition over time.

Bentonite's dry density (ρ_{db}) was defined on the basis of the dry density and the sand fraction where Bentonite fraction is responsible for the swelling pressure development and it was anticipated that the mixture of void ratio belongs only to bentonite [38-39] as follows:

$$\rho_{db} = \frac{(B/100)\rho_m G_{ss}}{G_{ss}(1 + w_m/100) - \rho_m(1 - B/100)} \tag{1}$$

where ρ_m (Mg/m^3)is the mixture density, B (%) is the bentonite content (in dry mass) in the mixture and is taken from the sand proportion S (%), w_m is the water content of the mixture, and G_{ss} is the specific gravity of sand (2.65).

TYPES

Bentonites are of four types such as sodium bentonite, calcium bentonite, potassium bentonite and aluminium bentonite.

1. Sodium Bentonite

Sodium bentonite is recommended as an absorbent aluminium silicate which is formed from the adjustment of volcanic ash where the chemical composition states that it has a higher percentage of sodium oxide (Na_2O) when compared to that of calcium oxide (CaO). The X-ray diffraction analysis [40-42] have evidently justified that it is impure clay showing a mineralogical composition of the prevailing montmorillonite. The sodium bentonite holds up active physical properties such as size of molecule which is low, large surface area, high cation exchange capacity, high absorption capacity and high swelling capacity. Owing to such properties, it is very beneficial in industries as well. The compound of aluminium silicate has an ability to procure the net negative charge for inducing large cation exchange capability. When the charge on the stratum of the clay is balanced by sodium ion, the characteristic properties are either acquired or enhanced respectively [43].

It is likewise due to the basic physical properties that the sodium bentonite is applied as a natural pozzolan for comparatively restoring the cement during the manufacture of cement materials. Because of the high reactivity to calcium hydroxide that generally liberate from cement hydrolysis during the generation of pozzolanic cementation bonds which establishes the continuous gain in strength and it also improves the microstructure of cemented materials. A large number of nucleation sites for the precipitation of the hydration product are propagated as it is spread all over the paste of cement [44]. The working ability gets improved, the pure culture and the induction in pozzolanic activity, because of such mechanism which is assigned to densification and homogenization of the paste. It also profoundly acts as filler for improvement in the packing adaptability of the cemented materials due to its fine composition.

2. Calcium Bentonite

Ca-bentonites were generally initialized by soda, which has varied proportions of soda. Alkali or soda stimulation with sodium carbonate (Na_2CO_3) was responsible for converting the calcium bentonites into sodium–calcium forms respectively. Gleason et al. [45] carried out the hydraulic conductivity test on a thin layer of calcium bentonite, which was percolated as well as saturated with the tap water. It was discovered after the trial that calcium bentonite was more permeable than the sodium bentonite. The impact of hydraulic performance was guided by nature and quality of bentonite [25-26]. When compared with sodium bentonite, the calcium bentonite has a specific property of smaller swelling capacity along with higher hydraulic conductivity and chemical resistance.

3. Potassium Bentonite

K-bentonites are an important tool in the epoch of palaeogeographic reconstructions and in the recent year it is aided for the knowledge of ancient explosive silicic volcanism which is preserved in the stratigraphic record [27]. The various physical properties such as pale yellowish and powdery, seldom sticky, palaeo-ash beds which generally varies in thickness between millimeters and few centimeters may sometimes even exceed to meter-scale [28-29].

4. Aluminium Bentonite

It has unit layer structure consisting of one Al^{3+} octahedral sheet which is placed between two Si^{4+} tetrahedral sheets. The permanent negative charges are seen in the bentonite because of the isomorphous replacement of Al^{3+} for Si^{4+} in the tetrahedral layer and Mg^{2+} for Al^{3+} in the octahedral layer. The cation Na^{+} and Ca^{2+} is counter balanced by such charges found in bentonite. These cations are exchangeable, which is also substituted by inorganic hydroxyl-metal polycations that are sustained into the layers irreversibly which later on concludes leading to the 'pillared' materials [30-33]. It was found that the quantities of hydroxyl-metal polycations generally include Al, Fe, Ti, Zn, etc and it was also implemented [34-40]. Hydroxyl-aluminium polycation ($[AlO_4Al_{12}(OH)_{24}(H_2O)_{12}]_n$) was employed randomly during those specific time period. Such pillared bentonites were enacted to adsorb metals, dyes, other pollutants as well as in industrial application [41]. It was also suggested by other physicists that such pillared bentonite showed outstanding ability for adsorbing humic acid (HA) [30].

METHODS

For the evidential description of flow behaviour of bentonite dispersion, the Herschel-bulkley model was applied for drilling the fluids [46].

$$\tau = \tau_y + K \cdot \gamma^n \tag{2}$$

where τ and τ_y are the shear stress and the yield stress respectively, K and n are consistency and flow index and γ is the shear rate.

The rheological parameters which basically include the yield stress, consistency and flow index was estimated from 3300 and 600 rpm dial readings on the basis of given formulas [47]

$$\tau_y = 0.511 R_3 \tag{3}$$

$$n = 3.322 \lg[(R_{600} - R_3)/(R_{300} - R_3)] \tag{4}$$

$$K = 0.511(R_{300} - R_3)/511^n \tag{5}$$

where R_3 is the 3 rpm dial reading, R_{300} is the 300 rpm dial reading and R_{600} is the 600 rpm dial reading respectively.

APPLICATIONS

1. Hydraulic Containment

With the hydraulic behavior dealt with an engineered barrier, it was found to be made of compact bentonite generally to limit the transfer of water and radioactive matters and was always found below the marginal level. Bentonite is originally unsaturated when it is compact at low water content where it experiences extremely high suctions and it is continuously wetted by water which is generated from the host formation. As swells has accompanied the vetting procedure, without taking under concern the thermal effect because of the passion that is given off from the waste canister, the barrier leading to water transport is coupled to mechanical phenomena which is related to swelling. Because of the intense large rigidity of the host formation that is granite for an instance, the bentonite vetting takes place in a condition of quasi constant volume.

From the given fact about compact bentonite, the hydraulic property can be employed as an engineered barrier material which is a basic cause for the designing of disposal system. This property was demonstrated and performed several times for briefing about the studies on it [48, 49].

Bentonite can be used for limiting the flow of liquid and transport of contaminants for a difference of hydraulic containment applications, such as groundwater cutoff walls, barriers for waste containment. It also includes landfills, wastewater ponds, manure lagoons, nuclear storage and the secondary containment in tank farms as well as seals in monitoring and water supply wells [50-54]. It was predominantly found that in all such applications, the sodium bentonite was employed, which proves that sodium cation is integrated with the exchange complex of the particles of bentonite. The basic advantages of sodium bentonite stems from the other appealing engineering properties, such as that of low hydraulic conductivity (k) to water (typically $k<10^{-10}$ m/s), and the existence of semipermeable membrane behaviour, or the restriction of solutes during migration is found to generally give advancement to the process namely hyperfiltration, chemico-osmosis, and reduced diffusive solute mass flux [55].

As the bentonite clay absorbs water or some organic molecules within the structural layers; it is observed to swell up, where it rolls up the properties of cationic exchange and the natural layer of silicates subsides the sub-microscopic particles [56-61]. The expansion of interlayer space is concerned with the swelling process which is very influential for the adsorption or intercalation of different organic molecules or species [62-63]. Bentonite layers bear a negative charge because of the isomorphic substitutions within the structure [64-66]. The layered structure and the existence of the net negative charge present in the minerals are basically due to the extraordinary properties such as the tendency to form colloids and their natural ion-exchange property [67].

The various chemical and physical parameters of such material are the consequence of the effects of diverse properties such as cation exchange capacity, layered structure, and dispensability in aqueous media and are found to be very significant for the process that occurs in nature and is used in various industrial applications [68]. Various organic dyes such as methylene blue, proflavine, acridine orange, thionine, crystal violet, methylene green, rhodamine B and rhodamine 6G are used as intercalated bentonite in the recent years [69-71]. The extension of clay mineral layered structure takes place as the bentonite is hung in the aqueous solution and it is also adaptable for ion exchange electrostatic interaction between the clay [68, 72, 73]. The p-interactions between positively charged aromatic dye entities and the clay oxygen planes lead to the optical phenomenon of interaction between clay and dye [74-75].

2. Oil Industry

There are a different range of applications related to the intercalation of bentonite with non-ionic surfactants in the oil industry [25-26]. The cationic surfactant based organic-bentonites are chemically less stable than the non-ionic ones where the more interlayer spacing is achieved by the of formation hydrogen bond between the oxide and the SiO_2 surface [27]. The interaction of bentonite with polypropylene glycol (PPG) which is likewise known as polypropylene oxide, was not studied briefly as such there was no current compromise on the swelling effect of it. PPG was applicable as a swelling agent for the synthesis of mesoporous materials [28], but yet it was assigned as a swelling inhibitor in oil and gas industry [25-26]. The non-ionic surfactants are swelling inhibitors, which was suggested on the basis of molecular dynamic investigation [26,29]. PPG are very well-known components for reducing the mixtures of concrete and it generally act in a suitable form during shrinkage as well [30].

CONCLUSION

Bentonite is natural clay which is formed as a result of weathering of volcanic ash generally in the presence of water. On the basis of the element which was dominant in composition usually the K, Na, Ca, and Al, bentonite was disintegrated into various types resembling different qualities. Due to the physical properties such as lower cost, environmental friendliness, stability, and reusability as well as its wide applications in pelletizing, fertilizers, seed coatings and effluent treatment, bentonite is highly preferable in the recent era [76].

REFERENCES

[1] S. Sang, J.Zhang, Q. Wu and Y. Liao, *Electrochim. Acta* 52, 7315 (2007).

[2] S. Meshram, R. Limaye, S. Ghodke and S. Nigam, *Chem. Eng. J.* 172, 1008 (2011).

[3] S.C. Motshekga, S.S. Ray, M.S. Onyango and M.N. Momba, *J. Hazard Mater.* 262, 439 (2013).

[4] N. Saelim, R. Magaraphan and T. Sreethwong, *Energ Convers. Manage.* 52, 2815 (2011).

[5] D.H. Solomon and D.G. Hawthorne, *Chemistry of Pigments and Fillers*, Wiley, New York (1983).

[6] H. van Olphen, *An Introduction to Clay Colloid Chemistry*, 2nd ed., Wiley, New York (1977).

[7] R.E. Grim, *Clay Mineralogy*, McGraw-Hill, New York (1968).

[8] J.M. Adams, J.A. Ballantine, S.H. Graham, R.J. Laub, J.H. Purnell, P.I. Reid, W.Y.M. Shaman and J.M. Thomas, *J. Catal.* 58, 238 (1979).

[9] S. Bohm, H.N. McMurray, S.M. Powell and D.A. Worsley, *Mater. Corros.* 52, 896 (2001).

[10] M.J. Loveridge, H.N. McMurray and D.A. Worsley, *Corros. Eng. Sci. Technol.* 41, 240 (2006).
[11] G. Williams, H.N. McMurray and D.A. Worsley, *J. Electrochem. Soc.* 149, B154 (2002).
[12] G. Williams and H.N. McMurray, in: G.S. Frankel, J.R. Scully, H.S. Isaacs and J.D. Sinclair (Eds.), Corrosion Science: *A Retrospective and Current Status in Honor of Robert P. Frankenthal, Electrochemical Society series*, Electrochemical Society Inc., Pennington New Jersey (2002).
[13] B.R.W. Hinton and L. Wilson, *Corros. Sci.* 29, 967 (1989).
[14] D.R. Arnott, B.R.W. Hinton and N. E. Ryan, *Mater. Perform.* 26, 211 (1987).
[15] A.J. Aldykiewicz, A.J. Davenport and H.S. Isaacs, *J. Electrochem. Soc.* 143, 147 (1996).
[16] H.N. McMurray, S.M. Powell and D.A. Worsley, *Corrosion* 55, 1040 (1999).
[17] S.Bohm, R. Greef, H.N. McMurray and D.A. Worsley, *J. Electrochem. Soc.* 147, 3286 (2000).
[18] K. Aramaki, *Corros. Sci.* 43, 1573 (2001).
[19] Y. Nakashima, *Clay Miner.* 41, 659 (2006).
[20] H.J. Herbert and H.C. Moog, *Eng. Geol.* 54, 55 (1999).
[21] C. Hurel and N. Marmier, *J. Radioanal. Nucl. Chem.* 284, 225 (2010).
[22] Lloret and M.V. Villar, *Phys. Chem. Earth* 32, 701 (2007).
[23] Imbert and M.V. Villar, *Appl. Clay Sci.* 32, 197 (2006).
[24] A.M. Tang, Y.J. Cui and T.T. Lee, *Appl. Clay Sci.* 41, 181 (2008).
[25] S.S. Agus, Swelling Pressure and wetting-drying curves of a highly compacted bentonite-sand mixture, in: T. Schanz (Eds.), *Unsaturated Soils: Experimental Studies*, Springer Berlin Heidelberg, New York, 93, 241 (2005)
[26] P. Delage, M.D. Howat and Y.J. Cui, *Eng. Geol.* 50, 31 (1998).
[27] H. Komine, *Eng. Geol.* 71, 265 (2004).
[28] H. Komine and N. Ogata, *Can. Geotech. J.* 31, 478 (1994).
[29] H. Komine and N. Ogata, *Can. Geotech. J.* 40, 460 (2003)
[30] H. Komine, K. Yasuhara and S. Murakami, *Can. Geotech. J.* 46, 177 (2009).
[31] R. Pusch, *Can. Geotech. J.* 19, 381 (1982).
[32] M.V. Villar and A. Lloret, *Appl. Clay Sci.* 39, 38 (2008).
[33] C.S. Tang, A.M. Tang, Y.J. Cui, P. Delage, C. Schroeder and E.D. Laure, Phys. Chem. Earth 36, 1857 (2011).
[34] L. Börgesson, O. Karnland and L.E. Johanneson, Eng. Geo. 41, 127 (1996)
[35] W.J. Cho, J.O. Lee and C.H. Kang, Ann Nucl Energy 27, 1271 (2000).
[36] J.O. Lee, J.G. Lim, I.M. Kang and S. Kwon, Eng. Geo. 129-130, 20 (2012).
[37] H.-J. Herbert, J. Kasbohm, H. Sprenger, A.M. Fernàndez and C. Reichelt, *Phys. Chem. Earth* 33, 327 (2008).
[38] D.A. Dixon, M.N. Gray and A.W. Thomas, Eng. Geo. 21, 247 (1985).
[39] J.O. Lee, W.J. Cho and K.S. Chun, J. Korean Nucl. Soc. 31, 139 (1999).
[40] C. Breen, G. Thompson and M. Webb, *J. Mater. Chem.* 9, 3159 (1999).
[41] J. L. Suter, P.V. Coveney, R.L. Anderson, H.C. Greenwell and S. Cliffe, *Energy Environ. Sci.* 4, 4572 (2011).
[42] Y.-H. Shen, *Chemosphere* 44, 989 (2001).

[43] B.-G. Park, W. Guo, X. Cui, J. Park and C.-S. Ha, *Micropor. Mesopor. Mater.* 66, 229 (2003).
[44] A.S. Bains, E.S. Boek, P.V. Coveney, S.J. Williams and M.V. Akbar, *Mol. Simulat.* 26, 101 (2001).
[45] J. Mora-Ruacho, R. Gettu and A. Aguado, *Cement Concrete Res.* 39, 141 (2009).
[46] S. B. Hamed and M. Belhadri, *J. Petrol. Sci. Eng.* 67, 84 (2009).
[47] J.N. Yan, Drilling Fluid Technology [M]. Petroleum Industry Press, Beijing (2003).
[48] H.M. Nachabe, *J. Irrig. Drain. Eng.* 121, 95 (1995).
[49] H. Komine, *Appl. Clay Sci.* 26, 13 (2004).
[50] M. Christman, C.H. Benson and T.B. Edil, *Groundw. Monit. Remediat.* 22, 104 (2002).
[51] P. Estornell and D.E. Daniel, *J. Geotech. Eng.* 118, 1592 (1992).
[52] L.S. Kajita, *Clay Clay Miner.* 45, 609 (1997).
[53] T. Katsumi, H. Ishimori and M. Onikata, R. Fukagawa, *Geotext. Geomembr.* 26, 14 (2008).
[54] J.A. Smith, S.L. Bartelt-Hunt and S.E. Burns, *J. Hazard. Mater.* 96, 91 (2003).
[55] M.A. Malusis, C.D. Shackelford and H.W. Olsen, *Eng. Geol.* 70, 235 (2003).
[56] F.G. Alabarse, R.V. Conceicao, N.M. Balzaretti, F. Schenato and A.M. Xavier, *Appl. Clay Sci.* 51, 202 (2011).
[57] S. Kaufhold and R. Dohrmann, *Appl. Clay Sci.* 51, 300 (2011).
[58] J.O. Lee, J.G. Lim, I.M. Kang and S. Kwon, *Eng. Geol.* 129-130, 20 (2012).
[59] P. Liu and L. Zhang, *Sep. Purif. Technol.* 58, 32 (2007).
[60] M.V. Villar, R. Gomez-Espina and L. Gutierrez-Nebot, *Appl. Clay Sci.* 65-66, 95 (2012).
[61] Q. Wang, X. Chang, D. Li, Z. Hu, R. Li and Q. He, *J. Hazard. Mater.* 186, 1076 (2011).
[62] J. Cenens and R.A. Schoonheydt, *Clay Clay Miner.* 36, 214 (1988).
[63] J. Bujdak, N. Iyi and T. Fujita, *Clay Miner.* 37, 121 (2002).
[64] H. Jia and C. Wang, *Chem. Eng. J.* 191, 202 (2012).
[65] A.M.L. Kraepiel, K. Keller and F.M.M. Morel, *J. Colloid Interface Sci.* 210, 43 (1999).
[66] F. Thomas, L.J. Michot, D. Vantelon, E. Montarges, B. Prelot, M. Cruchaudet and J.F. Delon, *Colloid Surface A* 159, 351 (1999).
[67] H.V. Olphen, An Introduction to Clay Colloid Chemistry: for Clay Technologies, Geologists, and Soil Scientists, 2nd ed., Wiley, New York (1977).
[68] Czímerová, A. Čeklovský and J. Bujdák, *Cent. Eur. J. Chem.* 7, 343 (2009).
[69] Z. Klika, H. Weissmannova, P. Capkova and M. Pospisil, *J. Colloid Interface Sci.* 275, 243 (2004).
[70] Gurses, C. Dogar, M. Yalcin, M. Acikyildiz, R. Bayrak and S. Karaca, *J. Hazard. Mater.* 131, 217 (2006).
[71] S. Yariv, D.K. Ghosh and L.G. Hepler, *J. Chem. Soc. Faraday T* 87, 1201 (1991).
[72] M. Holmboe, S. Wold and T. Petterson, *Parts A/B/C* 36, 1580 (2011).
[73] D.A. Laird, *Appl. Clay Sci.* 34, 74 (2006).
[74] J. Bujdak, N. Iyi and R. Sasai, *J. Phys. Chem. B* 108, 4470 (2004).
[75] Q. Chen, W.T. Kerk, A.M. Soutar and X.T. Zeng, *Appl. Clay Sci.* 44, 156 (2009).
[76] N. Boz, N. Degirmenbasi and D.M. Kalyon, *Appl. Cat. B: Environ.* 138-139, 236 (2013).

In: Bentonite
Editor: Ajay Kumar Mishra
ISBN: 978-1-63482-142-1

Chapter 6

UTILITY OF BENTONITE COMPOSITE FOR WASTEWATER TREATMENT

***Rohit Bhatia*[1,2], *Bharti Arora*[3], *Jitender Gaur*[4], *Eun Ha Choi*[5] and *Pankaj Attri*[5*]**

[1]Department of Science and Life, Satyawati College, University of Delhi, Delhi, India
[2]J & S Research and Innovations, New Delhi, India
[3]Department of Applied Sciences, ITM University, Haryana, India
[4]Council of Scientific and Industrial Research-National Physical Laboratory, Dr. K. S. Krishnan Marg, New Delhi, India
[5]Plasma Bioscience Research Center/Department of Electrical and Biological Physics, Kwangwoon University, Seoul, Korea

ABSTRACT

One of the major sources, which wastes the environment is the wastewater produced from the human activity due to enhanced discharge of toxic and persistent organic pollutants into the fresh water sources like rivers etc. Ground and surface waters contain, generally, impurities including organic pollutants such as pesticides, polychlorinated biphenyls (PCBs), halogenated organic solvents and polycyclic aromatic hydrocarbons (PAHs), discharged from industrial and wastewater treatment plants affect their appearance and may have adverse effects for consumers. Presence of large colloidal particle leads to clogging of filtering devices which increases cost of water purification. Bentonite is a kind of low-cost and easily obtained clay mineral. Dominant clay minerals in bentonites are smectites such as montmorillonite, beidellite, saponite, nontronite and hectronite. Natural Bentonite is effective adsorbent for cations but it shows lower affinity towards negative groups, like phosphate, due to the absence of reactive adsorption sites for anions in water. Bentonite clay can treat oil, sulfate, phosphate and is extremely effective at removing certain cationic components from wastewater and has a remarkable affinity for metals, particularly heavy metals in solution.

* Email: chem.pankaj@gmail.com

INTRODUCTION

The growing industrialization and extensive use of chemicals has increased the burden of unwanted pollutants in the drinking water. Each year Mediterranean olive-growing countries produce 10-30 million/m^3 of black and highly toxic olive mill to wastewater. Water in these areas contain impurities such as heavy metals, dyes, organic pollutants such as pesticides, polychlorinated biphenyls (PCBs), halogenated organic solvents, and polycyclic aromatic hydrocarbons (PAHs), etc. discharged from industrial and wastewater treatment plants, which affects their appearance and makes the water unsuitable for human consumption. Since it is impossible to prevent these chemicals from draining into the drinking water sources, the only way to prevent the further deterioration of water bodies is to develop efficient purifying technologies [1]. Natural clays are abundantly available low-cost natural resource which is nontoxic to ecosystem. Bentonite is such type of natural clay, with unique structural properties that makes it a valuable material for environmental application. It is one of the abundant clay minerals at the earth's surface, which can be used as a promising adsorbent for many toxic substances in water. This excellent capability is basically due to the presence of mineral montmorillonite. Bentonite with high surface area, prove to be an ineffective sorbent for non ionic organic compounds (NOC) in water due to hydrophilic nature of the mineral surfaces [2]. Over the recent years, various scientific communities have focused on bentonite to increase its adsorbent capacity and to remove other contaminants from drinking water other than metals is in progress. Therefore, scientists are paying keen attention to develop and characterize nanostructured clay composites because of their advantages in comparison to the traditional composites.

(1) Heavy Metals Removal Using Bentonite

Heavy metal contamination occurs in aqueous waste streams of many industries, such as metal plating facilities, mining operations, tanneries etc. Oil refineries, water mines, petrochemical plants, organics of synthesizing plants, coke-chemical etc. generate large amount of sewage containing variable amounts of harmful and toxic matters [3]. The adsorption of toxic metals, namely, Cd, Cr, Co, Cu, Pb, Mn and Ni has been studied predominantly. These being non-biodegradable, accumulate in living organisms and cause various diseases, disorders and environmental problems. Although they are essential for life in small concentrations, but excessive levels may be detrimental. Some of the processes including chemical precipitation, ion exchange, membrane separations (ultrafiltration, reverse osmosis, electrodialysis) and adsorption have been used for treatment of waste streams with metal contaminations [3]. The removal of heavy metal ions from industrial wastewaters using different adsorbents has acquired great interest in the past few days [4]. Some naturally occurring clay minerals have sorption capacity less than those of synthetic sorbents, but these materials could provide a cost-effective substitute for the removal of heavy metals. Bentonite has a remarkable affinity for heavy metals in the solution [5-9]. Magnetic nanoparticles and their composites serve as a convincing approach for the complete elimination of heavy metals from the drinking water [10]. It has been reported that the adsorption capacity of the clay iron oxide composite for the adsorption of Ni^{2+}, Cu^{2+}, Cd^{2+}, and Zn^{2+} metal ions from the aqueous solutions was better than the natural bentonite [11]. Also, the removal of nickel from aqueous solutions by the natural, treated and magnetically modified bentonite has been studied in batch adsorption-equilibrium experiments [3].

Among the different technologies, phosphorus removal by adsorption onto various materials has remained a subject of active investigation by the researchers worldwide. Bentonite clay has limited phosphorus adsorption capacity as the phosphate species are not favored by the typically negatively charged clay particles. Also, Al-Bentonite nanocomposite have been prepared and used for phosphorus removal. In particular, the study focused on assessing the impact of the particle size of the prepared Al-Betonite nanocomposite to remove phosphates from the solution. The Al-Bentonite nanocomposite samples have been prepared by mixing bentonite with selected Al-solutions, prepared with [OH]/[Al] ratios equal to 2.4. The adsorption results have shown significant phosphorus removal enhancement with the size reduction. The study has confirmed the high potential for using Al-Bentonite nanocomposite for phosphorus removal and the possibility of controlling the particle size of the adsorbent to achieve efficient phosphorus removal, combined with favorable separation following adsorption [12]. Also, cadmium (Cd) is one toxic heavy metal of particular environmental concern, because it can be introduced into and accumulated in the soil through agricultural application of sewage sludge, fertilizers, and/or through land disposal of Cd-contaminated municipal and industrial wastes. Several natural (e.g., natural zeolites, bentonite, metal oxides) and synthetic (e.g., synthetic zeolites, resins, metal phosphates and silicates, synthetic oxides/hydroxides/hydroxyl oxides) materials have been investigated as sorbents for heavy metal removal from the solutions, achieving different levels of success [13-17]. Due to high adsorption capacity for metal ions, clay minerals such as kaolinite [18], vermiculite [19] and sepiolite [20], have become the research hotpot for heavy metals removal from aqueous solution [21].

Hg(II) in industrial wastewaters prove to be a serious pollutant due to its toxic and bio-accumulative properties [22-23]. Recently, TiO_2 composites have been studied by preparing a series of compounds such as microcrystalline TiO_2 pillared clays, mixed Ti/Si, Ti/V oxides, nanocrystals of TiO_2 dispersed in the inorganic media [24-27]. Especially, TiO_2/ montmorillonite has been attracting much attention as a new low-cost type of photocatalyst, since its high surface area and superior properties accelerate photocatalytic reactions [26]. The adsorption and photocatalytic reduction of toxic Hg(II) in aqueous solutions at different temperatures have been investigated with synthesized TiO_2 nanoparticles and TiO_2/montmorillonite. It has been reported that the photocatalytic reduction of Hg(II) increases with the increasing temperature, and a decline in adsorption has also been observed for rise in temperature from 25 to 45°C, following the exothermicity of the adsorption process [28]. Lead is another potentially very hazardous metal to mammals and aquatic lives. The concentration of lead is increasing into the environment as a result of extensive and wide applications of mining, chemical, electroplating, petroleum refining, paper and pulp industries. It has been reported that the use of Polyacrylamide (PAA)-Bentonite (B) and zeolite (Z) composites and their Phy modifications can provide practicality and effectiveness for Pb^{2+} separation/removal procedures. It has been found that the chemical structure of the composites remained unchanged after five reuses and the storage foregoing [29]. The adsorption of Pb(II) from aqua solutions onto Unye (Turkey) bentonite in raw (RB) and manganese oxide-coated (MCB) forms has also been investigated [30].

It has been demonstrated by Zhou et al. [31] that Fe/Zr-montmorillonite can enable a promising application for removing Cr(VI) from the aqueous solutions. The equilibrium uptake of Cr(VI) has been found to be highly pH-dependent with an optimal pH range of 3.0-6.0 and the process of Cr(VI) adsorption onto pillared montmorillonites has been found out to be spontaneous and exothermic in nature. It has been predicted that the removal mechanisms of Cr(VI) by these pillared montmorillonites mainly involves electrostatic interaction and ion-exchange [31]. Similarly, Yuan group has investigated the removal mechanism of hexavalent chromium [Cr^{6+}] by montmorillonite-supported magnetite nanoparticles [32]. Bleiman and Mishael have designed Chitosan-montmorillonite composites to adsorb selenium from water and have compared its efficiency with widely used commercial adsorbent aluminum oxide (Al-oxide) and ferric oxide (Fe-oxide). The highest adsorption efficiency has been obtained in case of chitosan-montmorillonite composites. Selenium removal from well water (closed due to high selenium concentrations, 0.1 mg L^{-1}) by the composite, brought levels to below the WHO limit (0.01 mgL^{-1}) and has been found to be selective for selenium even in the presence of sulfur (13 mgL^{-1}) [33]. Zhao et al. [34] have investigated the sorption and desorption of Cu^{2+} on Bentonite-Polyacrylamide (BENT-PAAm) as the function of pH, ionic strength, adsorbent content, Cu(II) concentrations, and temperature. The sorption of Cu^{2+} on BENT-PAAm has been found to increase from about 9% to 97% at pH ranging from 2.4 to 7 and also the similar trend was observed with increasing temperature and decreasing ionic strength [34].

(2) Organic Contaminants Removal Using Bentonite

The adsorption of several organic contaminants in water such as pesticides, phenols, and chlorophenols has been reported recently in the literature [35-37]. Clays offer an attractive and inexpensive option for the removal of organic contaminants.

(a) Removal of Phenols and Cinnamic Acid

Phenols are being widely used in many industries such synthesis of resins, dyes, paint pharmaceuticals, perfumes, pesticides, tanning agents, solvents, or lubricating oils. The total production of phenol world over is estimated to be over 3 × 10^6 t/year. Phenol-containing water can easily be chlorinated resulting in the formation of toxic polychlorinated phenols which possess great environment problems, if released untreated. The long-term exposure of phenols to the human beings can lead to several problems like paralysis of central nervous system, kidneys malfunctioning and damage to lungs. However, the biodegradability of phenol in water is slow, about 90% in surface waters after 7 days with an aquatic toxicity (LC_{50}) of 12 mg/L (Daphnia magna, 48 h). So, there has been an increased focus on developing low cost catalytic system to remove phenol and its derivatives from water [38-39].

In another study, clay nanocomposite catalyst has been prepared for phenol oxidation and further the stability of hydrogen peroxide under different pH conditions has been studied. In pH range 3-4, H_2O_2 decomposed to least. It has been stated that at above pH 4, rapid decomposition of hydrogen peroxide occurred to produce molecular oxygen without the generation of noticeable amounts of hydroxyl radicals in the solution. Additionally, below pH 3, charged ferric ions remained as hydrated ferric ions, which not only led to hasten the decomposition of the peroxide but also led to competitive reaction for hydroxyl radicals' depletion [40-41]. Guo et al. [41] have studied the catalytic wet oxidation of phenol by hydrogen peroxide over bentonite nanocomposite with iron. The complete removal of aqueous phenol at 180 min at four H_2O_2 concentrations (25, 50, 100, and 600% of the stoichiometric amount) has been interpreted. Various intermediate products of phenol oxidation by hydrogen peroxide have been identified as acetic acid, catechol, and hydroquinone along with the measurable quantities of oxalic acid and benzoquinone, and malonic and formic acids in trace amounts. The hydroquinone, benzoquinone, and catechol have been identified during first stages of the phenol oxidation pathway. These by-products are as toxic as phenol and thus need to be removed as well. The proposed reaction Figure 1, shown below depicts the important interconnection between the presence of quinones and organic acids. The first step in oxidation of phenol is formation of intermediate compounds, catechol, hydroquinone, and benzoquinones. The later stage involves the cleavage of intermediates and the formation of carboxylic acids. Also, the direct decomposition of phenol to inorganic carbon during the ring-opening reaction has also been shown. About 30-40% of hydrogen peroxide homogeneously decayed in the liquid bulk, while the other portion decayed in contact with the clay catalyst [41]. The synthesis of bentonite nanocomposite using Al-Fe polymeric precursor has been reported by the Sanabria et al. [42]. They have prepared three different catalysts having mole ratios of Fe/metal used in the Al–Fe polymer synthesis (3.9, 10 and 23%). All catalyst used in this study successfully oxidized 100% of phenol after 2 h of reaction and out of which solid B2-Al-Fe was found to be best [42]. Guelou et al., [43] have oxidized phenol in semi-continuous reactor using bentonite nanocomposite with Al or mixed Al-Fe complexes with an Al-Fe atomic ratio equal to 9. They observed that the oxidation of phenol completed in less than 60 min and the rate of phenol conversion was unaffected by addition of hydrogen peroxide. Also, the time for total phenol oxidation increased with the decrease in the load of hydrogen peroxide solution [43].

Phenol degradation in sonophotocatalytic reactor using TiO_2/nanoclay nanocomposite has been reported recently. It has been reported that TiO_2/nanoclay nanocomposite effectively degraded phenol in 15 min under UV irradiation. The ideal loading concentration of nanocomposite was found to be 0.75 g/100 mL above which no noticeable increase in rate of phenol degradation was observed [44].

ZnO-bentonite nanocomposite photocatalyst has been used for removal of phenol under Continuous Stirred Tank Reactor (CSTR) system. The reactor walls were coated with thin film of composites and they showed nearly 70% of phenol removal from the simulated effluent at lower flow rate (10 ml min^{-1}) and higher pH, at a fixed nanocomposite loading. It has been suggested that removal of phenol follows 1st order reaction kinetics with Langmuir type of adsorption characteristics and phenol degradation was more pronounced under basic conditions rather than acidic ones mainly due to generation of phenoxide ion at pH 12. Further, the photocatalytic degradation of phenol resulted in the formation of adipic acid and 2,4,6-triphenoxy phenol through radical mechanism [45].

A clay nanocomposite with iron (6.1% (w/w) of Fe) for degradation of 2-chlorophenol (2-CP) has been reported using photo-Fenton conditions. On comparing the catalytic performance of the nanocomposite with other iron catalytic systems like zero valent iron (nZVI) and goethite, it has been reported that the rate of 2-CP degradation by nanocomposite was comparable to that of the nZVI and both catalysts, nanocomposite and nZVI, were found to be more efficient than goethite in 2-CP degradation in photo-Fenton conditions [46]. Again, degradation of 4-Chlorophenol using Al-Fe-, Al-Cu-and Al-Fe-Cu bentonite nanocomposite in catalytic wet peroxide oxidation method has been reported by Zhou et al., [47].

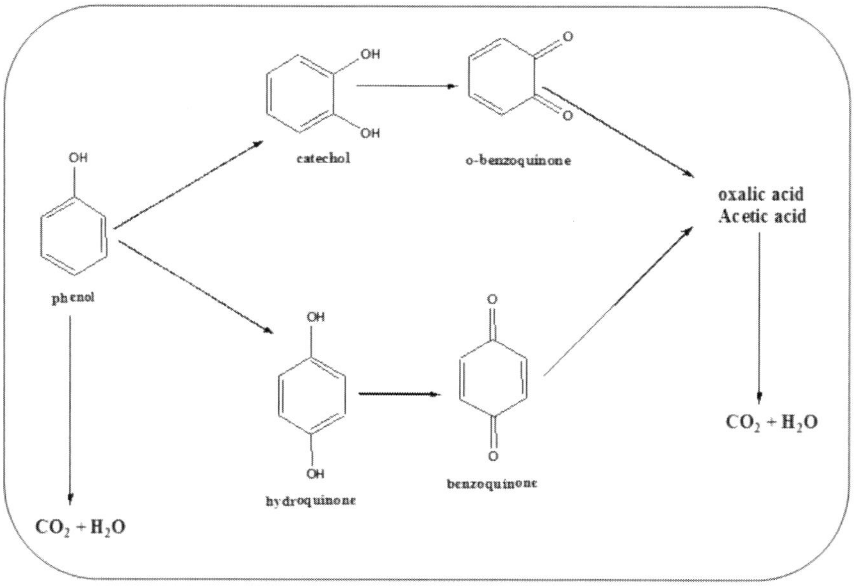

Figure 1. Proposed reaction scheme for phenol oxidation.

Chirchi et al. [48] have prepared 11 different catalysts, under different pH conditions and experimental procedures for degradation of 4-nitrophenol using various Fe-modified bentonite nanocomposite catalysts. It has been reported that PILC-(Al–Fe), a pillared sample made up by pillars of mixed Al and Fe oxides proved to be the best catalyst for the heterogeneous degradation of 4-nitrophenol [48]. The removal of two phenol derivatives 2,4,6-trinitrophenol (PA) and trichloro phenol (TCP) has been reported using nanocomposite prepared using poly-4-vinylpyridine-co-styrene (PVPcoS) and polydiallyl dimethylammonium chloride with clay. The PA is an anionic nitrophenol derivative with high soil mobility. The TCP is starting material for many herbicides and also commonly used in paper and pulp industry. Both PA and TCP have been shown to have higher binding affinity towards PVPcoS nanocomposite which can be attributed to van der Waals interactions. Among two pollutants, PA has been showing a higher affinity to the nanocomposite than TCP, indicating that electrostatic interactions also play a role besides van der Waals interactions. PA has been found to be removed by the composite within less than 4 min under wet composite condition, while in case of dry composite, time increased to 60 min [49].

Additionally, Fe-Bentonite nanocomposite have been used to remove the cinnamic acid from the olive oil industry effluent. It has been observed that pH 2.9 was the optimum for higher removal of cinnamic acid. Under heterogeneous condition, they have achieved 90% removal of cinnamic acid after 1h when Fe/cinnamic acid ratio = 2. While under homogenous conditions, they have observed that elimination of cinnamic acid increased with increase in Fe/cinnamic acid ratio up to value 10 which then became constant afterwards. The concentration of H_2O_2 also affects the rate of cinnamic acid removal, as maximum elimination of cinnamic acid under both heterogeneous and homogenous condition has been observed at H_2O_2/cinnamic acid molar ratio = 83 [50].

(b) Removal of Organic Dyes

The dyeing industry has been one of the biggest wastewater producing industries. The wastewater is characterized by high organic pollutant content, deep color, and significant impact on water quality. Worldwide nearly 10,000 types of dyes and pigments are produced every year. Further dye like azo dyes possesses additional risk in environment due to their non-biodegradable nature which makes their disposal more complex. Therefore, the considerable research in recent time has been devoted to use advance and cheap technology to get rid of dyes from effluents to make water potable [51-52].

The nanocomposite of chitosan, hexadecyl trimethyl ammonium bromide and clay has been reported for efficient removal of weak acid scarlet dye from aqueous solution. In comparison to nanocomposite, natural bentonite has shown very low adsorption capacity of weak acid scarlet. The optimal ratio of chitosan and CTAM for nanocomposite preparation has been found to be 1:10. The final removal efficiency was found to be higher than 85% and adsorption capacity reached 102.0 mg g^{-1} which was very higher in comparison to commercial activated charcoal [53].

The degradation of 4BS dye under visible-light irradiation using photocatalyst prepared by nitrogen and sulfur codoped TiO_2 clay nanocomposite has been reported. Red shift has been observed in case of nanocomposite as compared to pure TiO_2, which might be due to doping of N and S. The photocatalyst with a mole ratio of 1:4 of Ti:S has been found to perform best with the highest photodegradation rate. The UV-vis study of 4BS solution again indicates the complete loss of chromophore and partial decomposition of the naphthalene rings and benzene rings [54]. The degradation of azo dye X-3B has been reported using Fe-Bentonite and Al-Fe-Bentonite nanocomposite. Both nanocomposite Fe-Bentonite and Al/Fe-Bentonite possess high surface area 114.6 and 194.2 m^2/g which make them an effective catalyst. The optimum pH for reaction was found to be 3. Among both catalysts, Al-Fe Bentonite has shown better performance by discoloration of dye by 98% and removal of COD to be about 68% in 100 min. Meanwhile in Fe-Bentonite case, only 40% of discoloration has been observed in 100 min. The comparison of heterogeneous photo-Fenton and homogeneous photo-Fenton processes has shown that degradation rates of X-3B in the heterogeneous photo-Fenton seemed to be better than homogeneous photo-Fenton processes [55].

A Fe-Bentonite nanocatalyst has been used to degrade the 0.2 mM Orange II at nearly neutral pH 6.6. 100% degradation of 0.2 mM Orange II has been achieved in 90 min under the conditions as 1.0 g Fe-Bentonite/L nanocatalyst, 10 mM H_2O_2, and 8 W UV lamp. The major advantage of this system has been that there was no need to lower pH of industrial wastewater from neutral to about 3, which increased the cost of treatment [56]. The effect of pH of initial solution on the discoloration and mineralization of 0.2 mM Orange II dye has been studied using clay-Fe nanocomposite. And, it has been reported that best photocatalytic activity was observed at pH 3.0, and the same decreased with increase in as pH of initial solution. Under optimum conditions, clay-Fe nanocomposite achieved 100% discoloration and mineralization of dye in 60 and 120 min respectively. This property makes nanocomposite feasible for the photo-Fenton process even without adjusting pH of wastewater solution [57]. The nanocomposites have also been prepared by the pillaring technique. The reactor consisted of cylindrical stainless steel vessel whose inner wall was coated with Fe-Bentonite nanoparticles, a UV-C lamp (Philips 8 W, 254 nm) and H_2O_2 was added. The performance of nanocomposite has then been tested under various conditions like in dark, UV light alone, in absence of nanocomposite etc. Under dark conditions, very fast initial decrease in Orange II concentration due to the adsorption of the azo-dye on the surface of the nanocomposite has been observed. In absence of nanocomposite, significant oxidation of orange II due to generation of HO• from the photolysis of H_2O_2 has been reported. However, the best results have been obtained under photo-Fenton condition with complete degradation within 1 h. They attributed the degradation of Orange II due to adsorption on the Fe-Bentonite nanocomposite surface, oxidation by heterogeneous Photo-Fento reaction and oxidation by homogenous Photo-Fento reaction [58]. Further, comparison of various nanocomposites for degradation of Orange II has also been investigated. Bentonite nanocomposites with different Fe configurations (Fe polycation, α-Fe_2O_3 and α-FeOOH) using pillaring technique have been prepared and the rate of catalytic dye degradation has been outlined as; α-Fe_2O_3-Bentonite > hydroxyl-Fe-Bentonite > α-FeOOH-Bentonite. Additionally, the rate of dye mineralization has been reported to be α-Fe_2O_3-Bentonite = hydroxyl-Fe-Bentonite > α-FeOOH-Bentonite [59-60]. Feng et al. have used Fe-Bentonite nanocomposite for degradation of high concentration Orange II (2mM) and then compared it with laponite clay-Fe nanocomposite and iron oxide hydrated (FeOOH). They reported that by using 1g of catalyst 100% decoloration has been achieved in 120 min with laponite clay-Fe nanocomposite. While in case of Fe-Bentonite and iron oxide hydrated, it took longer (180 min) to degrade dye. The efficiency of both Fe-Bentonite and laponite clay-Fe nanocomposite have been found to be same in case of dye mineralization. The same trend has been observed with TOC removal [61].

Wan Ngah et al. [62] have prepared crosslinked chitosan/bentonite nanocomposites to adsorb azo dye, Tartrazine using epichlorohydrin as crosslinking agent. It has been further reported that sorption of tartrazine as a function of pH and rate of adsorption to be more at low pH. The adsorption capacity of nanocomposite has been found to be 294.1 mg/g [62]. Similarly Wang et al. [63] have used chitosan clay nanocomposite (molar ratio of 5:1) to eliminate Congo red from aqueous solution and reported dye adsorption to be dependent on the molar ratios of chitosan to clay, pH value of the dye solution and temperature. Further, nanocomposite were found to have flocculation ability which increased its dye removal efficiency [63]. The Fe-Bentonite nanocomposite system has been used for the degradation of another azo-dye, Acid Light Yellow G under heterogeneous catalyst conditions. The nanocomposite was found to have high photocatalytic activity resulting 100% discoloration and more than 65% TOC removal of 50 mg/L of dye [64]. The composite palygorskite-TiO_2 and halloysite-TiO_2 films with variable quantities of palygorskite and halloysite have been tested as photocatalysts in the photo-oxidation of Basic Blue 41 azo dye in water. These nanocomposite films have proven to be most promising photocatalysts and highly effective for decoloration of dyes [65].

Several studies have been conducted for non-azo dyes as well. Similarly, Hydroxy-Iron–Aluminum Bentonite (H-Fe-Al-B) nanocomposite has been reported to degrade the Azocarmine B. The optimum condition for the degradation of azocarmine B have been found to be, dye concentration of 143.7 mg/L, pH 4.2 and H_2O_2 concentration as 17.7 mM, respectively. The degradation rate was found to be 99.5% under these conditions [66]. Bentonite-FeCo nanocomposite hybrid hydrogel has been synthesized using ultrasound assisted in situ emulsion polymerization to remove cationic dye, Crystal violet (CV). The rate of desorption of dye was found to be more efficient at higher pH and temperature 35 ^{o}C. It has been suggested that this could be due to dissociation of COO^- ions at higher pH [67]. Magnetic TiO_2/clay/Fe_3O_4 nanocomposite have been used to degrade Methylene Blue dye

(MB). The TEM studies have shown that TiO_2 of size 10-20 nm were embedded in layer of clay, while the Fe_3O_4 of size 45-60 nm was deposited on clay surface. The TiO_2/clay/Fe_3O_4 nanocomposite has been found to successfully decompose nearly 94% of the initial MB after 80 min of UV irradiation. The nanocomposite has good recyclability, it retained its activity till 6 cycles with a high degradation rate [68].

The polyepicholorohydrin-dimethylamine and bentonite nanocomposite have been prepared and their ability to remove different non-ionic dyes (Disperse Blue SBL and Vat Scarlet R) and anionic dyes (Reactive Violet K-3R and Acid Dark Blue 2G) have also been explored. It has been delineated that efficiency dye removal was significantly affected by solution pH, presence of salt and surfactant. And, the amount of nanocomposite for 95% dye removal in 5 L dye solution, with a concentration of 50 mg/L was found to be 378.0 g, 126.5 g , 9.7 g and 15.5 g for DB SBL, VS R RV K-3R, ADB 2G, respectively [69]. In another study, adsorption capacity of 435.0 mg/g of Malachite Green using same nanocomposite has been achieved [62,70]. The Fe-Ni-clay nanocomposite has been reported to remove basic Magenta dye from aqueous solution and the presence of spherical Fe-Ni nanoparticles of size 30–40 nm were found to intercalate onto clay layers. Also, the nanocomposite was found to be effective even after 12 cycles. Both nanocomposites showed appreciable BM removal capability (222-750 mg L^{-1}) per gm of nanocomposites [71].

(c) Removal of Herbicides & Pesticides

Atrazine (2-chloro-4-ethylamino-6-isopropylamino-s-triazine) as shown in Figure 2(a) has been among the most widely applied herbicides and seems to be a common contaminant of the ground water.

The most commonly used technology for atrazine removal from water has been the adsorption by granular activated carbon (GAC) [72]. Atrazine removal from water by two polycations pre-adsorbed on montmorillonite has been compared with GAC. It has been stated that the most suitable composite poly (4-vinylpyridine-co-styrene)-montmorillonite (PVP-co-S90%-mont.) removed 90-99% of atrazine (0.5-28 ppm) within 20-40 min at 0.367% w/w. In the presence of dissolved organic matter (DOM; 3.7 ppm), the efficiency of the GAC filter to remove atrazine decreased significantly (68-52% removal), whereas the corresponding efficiency of the PVP-co-S90%-mont. filter was only slightly influenced by DOM. At lower atrazine concentration (7 ppb,) PVP-co-S90%-mont. filter reduced even after 3000 pore volumes the emerging atrazine concentration below 3 ppb (USEPA standard). Thus, PVP-co-S90%-mont. composite has been manifested as a new efficient material for the removal of atrazine from water [73].

Fluometuron (N,N-dimethyl-N'-{3-(trifluoromethyl)phenyl} urea) as shown in Figure 2(b), is a phenyl urea herbicide, which has been widely used to control grass and broad-leaved weeds in cotton and olive crops. It has the tendency to diffuse through permeable and sandy soils causing groundwater contamination. Interaction between the naturally occurring polycation spermine (SPERM) and Na-exchanged SWy-2 Wyoming montmorillonite (Na-SWy-2 Mt) and between spermine exchanged montmorillonite (Mt-SPERM) and the herbicide fluometuron has been well studied. It has been evidenced earlier that intercalation of spermine into Na-SWy-2 increased the thermal stability of the organic cations and affected the dehydroxylation temperature of the clay mineral. Also, the involvement of N-H and C-H groups of spermine in the interaction with the clay mineral has been predicted. Mt-SPERM has thus been shown to have pronounced affinity for the herbicide fluometuron [74].

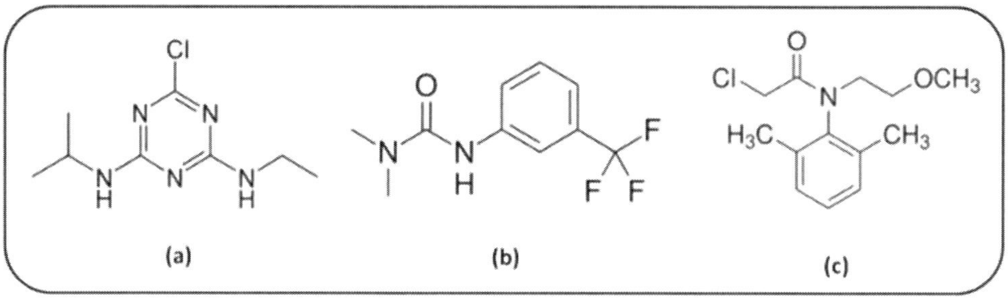

Figure 2. Structures of (a) Atrazine, (b) Fluometuron, and (c) Dimethachlor.

Polyethersulfone/organically modified montmorillonite nanocomposite membranes have been reported for removal of pesticide from water. The nanocomposites have been prepared by a combination of solution dispersion and wet-phase inversion methods. Further, Ghaemi et al. [75] have reported that the addition of different concentrations of clay to the casting solution resulted in formation of membranes with a thinner skin layer and a smaller mean pore size. The removal of pesticide by nanocomposite membranes, by solute adsorption on the membrane surface and electrostatic charge repulsion between charged pesticides with similar charges on the membrane surface has also been attributed [75].

TiO_2/clay nanocomposites have been used to remove a chloroacetanilide herbicide (Dimethachlor) (as shown in Figure 2(c)) from water. The montmorillonite (Z)-TiO_2 samples containing 15, 30 and 55 wt% TiO_2 were labelled as ZT_a, ZT_b and ZT_c. Studies have shown that nanocomposite has small crystal sizes, large porosity and specific surface areas which make it better a photocatalyst. It has been established that degradation of dimethachlor followed pseudo-first-order kinetics according to the Langmuir–Hinshelwood mode. The % adsorption of dimethachlor on different nanocomposite followed the order: $ZT_a > ZT_c \approx ZT_b > ZT_a > TiO_2$ (blank) $\approx$ TiO_2 P25 and order of photocatalytic activity seemed to be, $ZT_c > ZT_b > ZT_a$ [76].

(d) Removal of Endocrine Disruptors

TiO_2/clay nanocomposite have also been reported for adsorption-photocatalytic degradation of various endocrine disruptors (di-*n*-butyl phthalate, diethyl phthalate, dimethyl phthalate and bisphenol-A). The nanocomposite has been synthesised using hydrothermal method to improve the crystallinity of the TiO_2 resulting in a better photocatalyst activity. It has been reported that the presence of hydrophobic interlayer in TiO_2/clay nanocomposite leads to enhanced photocatalytic degradation [77].

(3) Winery Industries Wastewater Treatment Using Bentonite

Wineries have been among another important agricultural industry which produce large quantity organically laden wastewater (1000-3000 L per ton of grapes). This wastewater has high concentrations of organic material and nutrients, high acidity and large variations in seasonal flow production. Due to high concentration of organic matter, suspended solids and sodium adsorption ratio (SAR) make water inadequate for disposal in common sewage systems [78-82]. Rytwo et al. [83] have reported the use of clay-polymer nanocomposites for clarification of winery wastewater. The nanocomposites have been prepared using cationic polymer poly-DADMAC (Polydiallyldimethylammonium chloride) and chitosan with clay which leads to winery effluents clarification due to neutralization of the colloids [83].

CONCLUSION

It is clear that natural clay and its composites are capable of removing contaminants ranging from metals to priority pollutants in contaminated drinking water. In most of the cases, they proved to be better or comparable with the existing commercial filter materials, adsorbents, and conventional methods used for decontamination of drinking water. They can effectively remove toxic heavy metals, organic contaminants including dyes, phenols, acids, herbicides and pesticides. Bentonite being natural clay makes it composite a low-cost, green, nontoxic adsorbent which can be used for removal of different contaminants from water and making the availability of clean and pure drinking water for developed and developing nations.

REFERENCES

[1] R. Srinivasan, Advances in Materials Science and Engineering, Vol. 2011, p.1, 2011.
[2] Y.H. Shen, *Water Research,* Vol. 36, p. 1107, 2002.
[3] J. Veres and Z. Orolinova, *Acta Montanistica Slovaca,* 2009, Vol. 14, p. 152.
[4] S.E. Bailey, T.J. Olin, R.M., Bricka, and D.D. Adrian, *Wat. Res.,* Vol. 33, p. 2469, 1999.
[5] M.C. Jesse and A.H., American Water Werks Association, Inc., 3rd eds. McGraw Hill Book, 1971.
[6] C.D. Cancela, E.R. Tabooada, and F.S. Rasero, *Soil Science,* Vol. 150, p. 836,1990.
[7] D.A. Laird, P.Y. Yeen, W.C. Koskinen, T.R. Steinheimer, and R.H. Dowdy, *Environ. Sci. Technol.,* Vol. 28, p. 1054, 1994.
[8] M. Jeffery and P.E. Harris, Intercontinental Landfill Research Symposium, Lulea University of Technology, Lulea, Sweden, p. 16, 2000.
[9] A.M. Abdelaal, M.S. EL-Salmawy, and A. A. Abdelrahman, The Fifth Egyptian Syrian Conference In Chemical and Petroleum Engineering, Faculty of Petroleum & Mining Engineering, Suez, Egypt, Vol. 1, p. 300, 2003.
[10] T. Szabo, A. Bakandritsos, V. Tzitzios, S. Papp, L. Korosi, G. Galbac, K. Musabekov, D. Bolatova, D. Petridis, and I. Dekany, *Nanotechnology,* Vol. 18, p. 285602, 2007.
[11] L.C.A. Oliveira, R.V.R.A. Rios, J.D. Fabris, K. Sapag, V.K. Garg, and R.M. Lago, *Appl. Clay Sci.,* Vol. 22, p. 169, 2003.
[12] M. EL-Sergany and A. Shanableh, Asia Pacific Conference on Environmental Science and Technology Advances in Biomedical Engineering, Vol.6, p. 323, 2012.
[13] E. Erdem, N. Karapinar and R Donat, *Journal of Colloid and Interface Science,* Vol. 280, p. 309, 2004.
[14] O. Abollino, M. Aceto, M. Malandrino, C. Sarzanini, and E. Mentatsi, Water Research, Vol. 37, p. 1619, 2003.

[15] K. A. Matis, A. I. Zouboulis, G. P. Gallios, T. Erwe, and C. Blocher, *Chemosphere,* Vol. 55, p. 65, 2004.

[16] D. Zamboulis, S. I. Pataroudi, A. I. Zouboulis, and K. A. Matis, *Desalination,* Vol. 162, p. 159, 2004.

[17] Xu and L. Axe, Journal of Colloid and Interface Science, Vol. 282, p. 11, 2005.

[18] M. Alkan, B. Kalay, M. Dogan, and O. Demirbas, *J. Hazard. Mater,* Vol. 153, p. 867, 2008.

[19] M. Malandrino, O. Abollino, A. Giacomino, M. Aceto, and E. Mentasti, *J. Colloid Interface Sci.,* Vol. 299, p. 537, 2006 .

[20] M. Dogan, Y. Turhan, M. Alkan, H. Namli, P. Turan, and O. Demirbas, *Desalination,* Vol. 230, p. 247, 2008.

[21] P. Wu, W. Wu, S. Li, N. Xing, N. Zhu, P. Li, J. Wu, C. Yang, and Z. Dang, *Journal of Hazardous Materials,* Vol. 169, p. 824, 2009.

[22] R. Herrero, P. Lodeiro, C. Rey-Castro, T. Vilarino, and D. S. Vicente, *Water Res.,* Vol. 39, p. 3199, 2005.

[23] X. L. Wang, S. O. Pehkonen, A. K. Ray, Electrochim. Acta, Vol. 49, p. 1435, 2004.

[24] V. Belessi, D. Lambropoulou, I. Konstantinou, A. Katsoulidis, P. Pomonis, D. Petridis, and T. Albanis, *Appl. Catal. B: Environ.,* Vol. 73, p. 292, 2007.

[25] X. Ding, T. An, G. Li, S. Zhang, J. Chen, J. Yuan, J. Zhao, H. Chen, G. Sheng, and J. Fu, *J. Colloid Interface Sci.,* Vol. 320, p. 501, 2008.

[26] R. Kun, K. Mogyorosi, and I. Dekany, *Appl. Clay Sci.,* Vol. 32, p. 99, 2006 .

[27] J. Yang, J. Zhang, L. Zhu, S. Chen, Y. Zhang, Y. Tang, Y. Zhua, and Y. Li, *J. Hazard. Mater. B,* Vol. 137, p. 952, 2006.

[28] B.L. Dou, V. Dupont, W.G. Pan, and B.B. Chen, *Chemical Engineering Journal,* Vol. 166, p. 631, 2011.

[29] U. Ulusoy and S. Şimsek, *Journal of Hazardous Materials,* Vol. 127, p. 163, 2005.

[30] E. Eren, B. Afsin, and Y. Onal, *Journal of Hazardous Materials,* Vol. 161, p. 677, 2009.

[31] J. Zhou, P. Wu, Z. Dang, N. Zhu, P. Li, J. Wu, and X. Wang, *Chemical Engineering Journal,* Vol. 162, p. 1035, 2010.

[32] P. Yuan, M. Fan, and D. Yang et al., *Journal of Hazardous Materials,* Vol. 166, p. 821, 2009.

[33] N. Bleiman and Y. G. Mishael, *Journal of Hazardous Materials,* Vol. 183, p. 590, 2010.

[34] G. Zhao, H. Zhang, and Q. Fan et al., *Journal of Hazardous Materials,* Vol. 173, p. 661, 2010.

[35] J. M. Rodriguez, A. J. Lopez, and S. Bruque, Clays & Clay Minerals, Vol. 36, p. 284, 1988.

[36] T. G. Danis, T. A. Albanis, D. E. Petrakis, and P. J. Pomonis, *Water Research,* Vol. 32, p. 295, 1998.

[37] K. Konstantinou, T. A. Albanis, D. E. Petrakis, and P. J. Pomonis, *Water Research,* vol. 34, p. 3123, 2000.

[38] C.H. Chiou and R.S. Juang, *J. Hazard. Mater.,* Vol. 149, p. 1, 2007.

[39] B. Iurascu, I . Siminiceanu, D. Vione, M.A. Vicente, and A. Gil, *Water Res.,* Vol. 43, p. 1313, 2009.

[40] J.M. Tatibouet, E. Guelou, and J. Fournier, *Top Catal,* Vol. 33, p. 225, 2005.

[41] J. Guo and M. Al-Dahhan, *Ind. Eng. Chem. Res.*, Vol. 42, p. 2450, 2003.
[42] N. Sanabria, A. A. lvarez, R. Molina, and S. Moreno, *Catalysis Today,* Vol. 133-135, p. 530, 2008.
[43] E. Guelou, J. Barrault, J. Fournier, and J.M. Tatibouet, *Appl. Catal. B Environ,* Vol. 44, p. 1, 2003.
[44] S. Ghodke, S. Sonawane, R. Gaikawad, and K. C. Mohite, *Can. J. Chem. Eng.,* Vol. 90, p. 1153, 2012.
[45] S. Meshram, R. Limaye, S. Ghodke, S. Nigam, S. Sonawane, and R. Chikate, *Chemical Engineering Journal,* Vol. 172, p. 1008, 2011.
[46] M.A. De Leon, M. Sergio, J. Bussi, G.B. Ortiz de la Plata, A.E. Cassano, and O.M. Alfano, *Environ Sci Pollut Res.,* 2014 (doi: 10.1007/s11356-014-2681-6).
[47] S. Zhou, C. Zhang, X. Hu, Y. Wang, R. Xu, C. Xia, H. Zhang, and Z. Song, *Applied Clay Science,* Vol. 95, p. 275, 2014.
[48] L. Chirchi and A. Ghorbel, *Appl. Clay Sci.,* Vol. 21, p. 271, 2002.
[49] R. Ganigar, G. Rytwo , Y. Gonen, A. Radian , and Y.G. Mishael, *Applied Clay Science,* Vol. 49, p. 311, 2010.
[50] D. Tabet, M. Saidi, M. Houari, P. Pichat, and H. Khalaf, *J. Environ. Manage,* Vol. 80, p. 342, 2006.
[51] Q. Chen, P. Wu, Y. Li, N. Zhu, and Z. Dang, J Hazard Mater, Vol. 168, p. 901, 2009.
[52] P.A. Carneiro, R.F. Pupo Nogueira, and M.V.B. Zanoni, Dyes Pigm, Vol. 74, p.127, 2007.
[53] J. Guo, S. Chen , L. Liu, B. Li, P. Yang, L. Zhang, and Y. Feng. *Journal of Colloid and Interface Science,* Vol. 382, p. 61, 2012.
[54] G. Zhang, X. Ding, Y. Hu, B. Huang, X. Zhang, X. Qin, J. Zhou, and J. Xie. *J. Phys. Chem.,* C, Vol. 112, p. 17994, 2008.
[55] Y. Li, Y. Lu, and X. Zhu, *Journal of Hazardous Materials,* B, Vol. 132, p. 196, 2006.
[56] J. Feng, R.S.K. Wong, X. Hu, and P.L. Yue. Catalysis Today, Vol. 98, p. 441, 2004.
[57] J. Feng, X. Hu, and P.L. Yue, *Water Research,* Vol. 40, p. 641, 2006.
[58] J. Feng, P.L. Yue, X. Hu, and R.S.K. Wong, US20060076299A1, 2006.
[59] J. Chen and L. Zhua, *Separation and Purification Technology,* Vol. 67, p. 282, 2009.
[60] J. Chen and L. Zhu, *Journal of Photochemistry and Photobiology,* A: Chemistry, Vol. 188, p. 56, 2007.
[61] J. Feng, X. Hu, P.L. Yue, S. Qiao, *Separation and Purification Technology,* Vol. 67, p. 213, 2009.
[62] W.S. Wan Ngah, N.F.M. Ariff, and M.A.K.M Hanafiah, Water, *Air and Soil Pollution,* Vol. 206, p. 225, 2010.
[63] L. Wang and A. Wang, *Journal of Hazardous Materials,* Vol. 147, p. 979, 2007.
[64] J. Chen and L. Zhu, *Catalysis Today,* Vol. 126, p. 463, 2007.
[65] D. Panagiotaras, D. Papoulis, and E. Stathatos, *International Journal of Environmental, Earth Science and Engineering,* Vol. 8, p. 175, 2014.
[66] T. Xu, Y. Liu, F. Ge, L. Liu, and Y. Ouyang, *Applied Surface Science,* Vol. 280, p. 926, 2013.
[67] S.R. Shirsath, A.P. Hage, M. Zhou, S.H. Sonawane, and M.A. Kumar, *Desalination,* Vol. 281, p. 429, 2011.
[68] Ping Zhang, Zunli Mo, Lijuan Han, Xiaobo Zhu, Bo Wang, and Chun Zhang, *Ind. Eng. Chem. Res.,* Vol. 53, p. 8057, 2014.

[69] Q. Li, Q.Y. Yue, H.J. Sun, Y. Su, and B.Y. Gao, *Journal of Environmental Management,* Vol. 91, p. 1601, 2010.

[70] W.S. Wan Ngah, N.F.M. Ariff, and M.A.K.M Hanafiah, *Clean,* Vol. 38, p. 394, 2010.

[71] B.S. Kadu and R.C. Chikate, *Chemical Engineering Journal,* Vol. 228, p. 308, 2013.

[72] Z. Yue, J. Economy, K. Rajagopalan, G. Bordson, M. Piwoni, L. Ding, V.L. Snoeyink, and B.J. Marinas, *J. Mater. Chem.,* Vol. 16, p. 3375, 2006.

[73] D. Zadaka, S. Nir, A. Radian, and Y.G. Mishael, *Water Research,* p. 1, 2008.

[74] B. Gamiz, R. Celis, M.C. Hermosin, J. Cornejo, and C.T. Johnston, *Applied clay Science,* Vol. 58, p. 8, 2012.

[75] N. Ghaemi, S. S. Madaenia, A. Alizadeh, H. Rajabi, and P. Daraei, *Journal of Membrane Science,* Vol. 382, p. 135, 2011.

[76] V. Belessi, D. Lambropoulou, I. Konstantinou , A. Katsoulidis , P. Pomonis , D. Petridis, and T. Albanis, Applied Catalysis B: *Environmental,* Vol. 73, p. 292, 2007.

[77] C. Ooka, H. Yoshida, M. Horio, K. Suzuki, and T. Hattori, Applied Catalysis B: *Environmental,* Vol. 41, p. 313, 2003.

[78] G. Rytwo, *World Applied Sciences Journal,* Vol. 19, p. 874, 2012.

[79] Jail, F. Boukhoubza, A. Nejmeddine, S. Sayadi, and L. Hassani, *Journal of Hazardous Materials,* Vol. 176, p. 893, 2010.

[80] J. Chapman, Winery Wastewater Handbook, Winetitles, Adelaide, Australia, 2001

[81] F. Masi, G. Conte, N. Martinuzzi, and B. Pucci, Proceedings of the IWA 8th *International Conference on Wetland Systems for Water Pollution Control,* Vol. 1, p. 274, 2002.

[82] Brito, J. Peixoto, and J. M. Oliveira et al., Utilization of By-Products and Treatment of Waste in the Food Industry, V. Oreopoulou and W. Russ, Eds., p. 109, *Springer,* 2007.

[83] G. Rytwo, R. Lavi , Y. Rytwo , H. Monchase , S. Dultz , T. N. Konig, *Science of the Total Environment,* Vol. 442, p. 134, 2013.

In: Bentonite
Editor: Ajay Kumar Mishra
ISBN: 978-1-63482-142-1

Chapter 7

BENTONITE: A VERSETILE CLAY

***Pradeep Pratap Singh*[1,*] *and Ambika*[2]**
[1]Department of Chemistry, Swami Shraddhanand College,
University of Delhi, Delhi
[2]Department of Chemistry, Hans Raj College,
University of Delhi, Delhi

ABSTRACT

Bentonites are colloidal clays composed of clay mineral montmorillonite, resulting from the decomposition of volcanic ash. They are highly water absorbent with high plasticity, and have high swelling and shrinkage characteristics. The properties of bentonite are derived from the crystal structure of the smectite group, which is an octahedral alumina sheet between two tetrahedral silica sheets. Variations in interstitial water and exchangeable cations in the interlayer space affect the properties of bentonite. Bentonites are used in papermaking technology, cut-off walls, wastewater purification and in different aspects of nano-technology. Bentonite has also found medical applications and has been used in the form of a gel, as a bulk laxative and also as a base in many dermatological formulations. They are also used as feed and food additives. Bentonites have been found to reduce levels of radiocaesium in animals and their products. This article includes the versatile aspects of bentonites and their applications.

INTRODUCTION

Bentonite was named after Fort Benton (Wyoming, USA), the locality where it was first found. Most high-grade natural sodium bentonite is produced from the western United States. Mixed sodium/calcium bentonite is mined in Greece, Australia, India, Russia, and Ukraine (Figure 1). Calcium bentonite is mined in United States, Germany, Greece, Turkey, India, and China. Bentonite is a rock made up of highly colloidal and plastic clays composed mainly of montmorillonite, a clay mineral of the smectite group [1]. In addition to montmorillonite, it may also contain feldspar, biotite, kaolinite, illite, cristobalite, pyroxene, zircon, and crystalline quartz [2]. It was geologically formed from volcanic ash approximately 100 million years ago. Bentonite was formed when temperature and pressure were applied to the volcanic ash and lava deposited on the bottoms of seas and lakes for a very long period (several million years) that resulted in change of the molecular structure of the minerals (Figure 2).

Bentonites are white, light weight rock deposits composed mostly of salts of hydrated aluminosilicates of sodium (Na), potassium (K), calcium (Ca), and occasionally iron, magnesium, zinc, nickel, etc. These bentonites have a high negative charge and are balanced by cations such as Mg, K, and Na located in the cavities; therefore, they act as inert material due to their neutral pH or slightly alkaline nature [3].

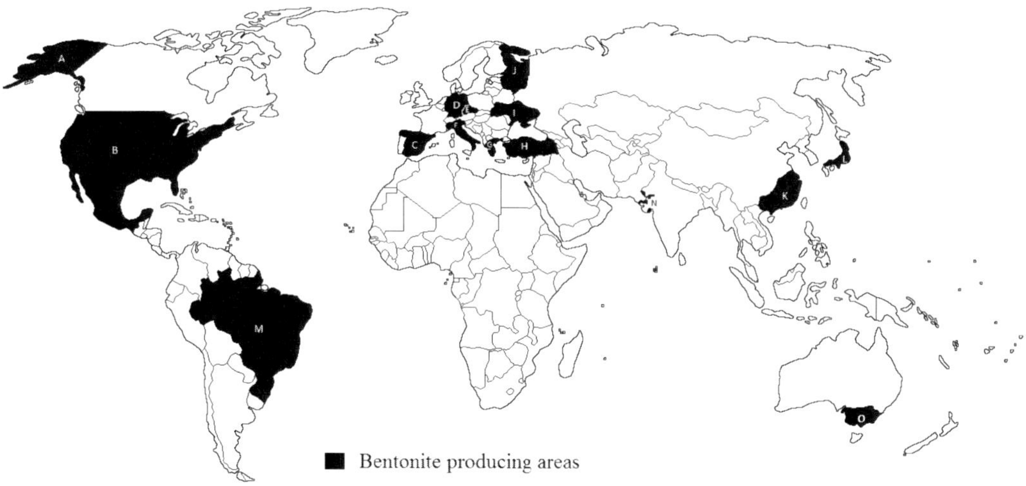

Figure 1. Geographical distribution of bentonite.

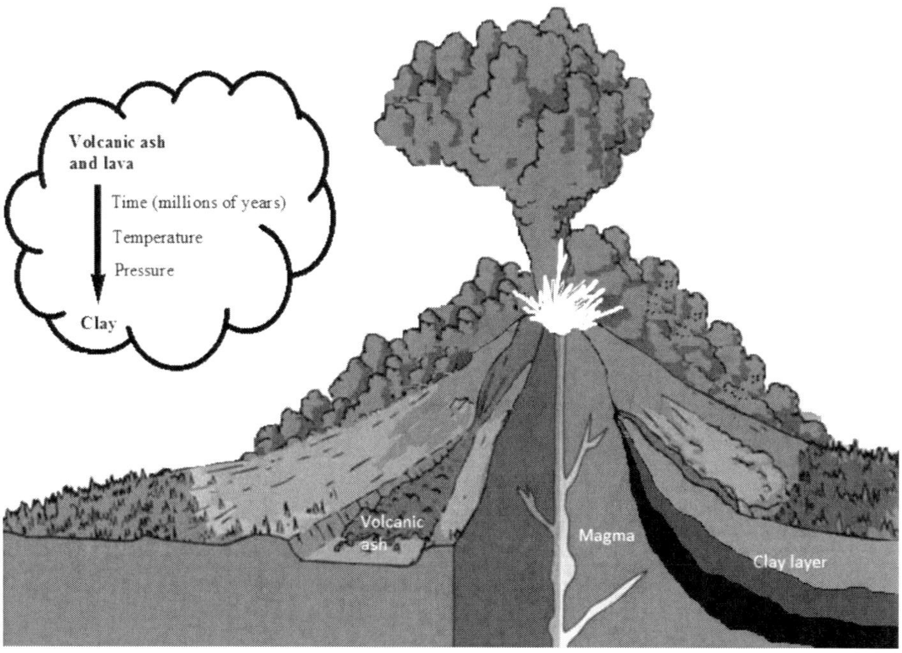

Figure 2. Formation of bentonite clay from volcanic eruption.

The properties of bentonite are derived from the crystal structure of the smectite group, which is an octahedral alumina sheet between two tetrahedral silica sheets. Variations in interstitial water and exchangeable cations in the interlayer space affect the properties of bentonite and hence the commercial uses of the different types of bentonite. Their properties can vary considerably depending on geological origin and any post-extraction modification [4].

Bentonites are clays with strong colloidal properties that absorb water rapidly, which results in swelling and a manifold increase in volume, giving rise to a thixotropic, gelatinous substance [5]. This, together with several other properties makes natural sodium bentonite the ideal material for the safe sealing of reinforced concrete structures. Its commercial importance depends more on its physico-chemical properties rather than on its chemical composition. It is important due to its excellent plasticity and lubricity, high dry-bonding strength, high shear and compressive strength, low permeability and low compressibility. It is valued in foundry sand binding [6], drilling mud [7], in decolourizing oils [8,9], in the manufacture of catalysts [10] and as waterproofing and sealing agent in civil engineering works [11] and in pharmaceuticals [12,13]. Numerous more minor applications have also been reported, including some in food production and animal nutrition [14].

Bentonite in India

In India Rajasthan (80%) and Gujarat (18%), are the major bentonites producing states, while minor amounts are produced in Tamil Nadu, Jharkhand and Jammu & Kashmir. Among these states, Bhavnagar and Kachchh districts of Gujarat and Barmer district of Rajasthan are the major producers of bentonite. The sodium bentonite mined in Rajasthan tends to be of lower quality and is used as foundry sand. It is exported in unprocessed (crude) and processed (including activated) forms as well [15].

Structural Properties

Bentonite is naturally occurring 2:1 type clay, mainly composed of montmorillonite. The basic structure of bentonite consists of two silica tetrahedral sheets with an intermediate aluminium octahedral sheet (Figure 3) [16,17]. The charge between octahedral and tetrahedral sheets is not balanced due to the isomorphous substitution of Al^{3+} for Si^{4+} in the tetrahedral sheets and generally Mg^{2+} and Al^{3+} in the octahedral sheets [18]. The substitution of these ions of lower valency induces a permanent negative charge in the lattice structure. The negative charge is balanced by treating it with cations such as sodium, calcium and magnesium. These cations are exchangeable, and depending on the exchangeable cation the bentonite is named as sodium bentonite or calcium bentonite.

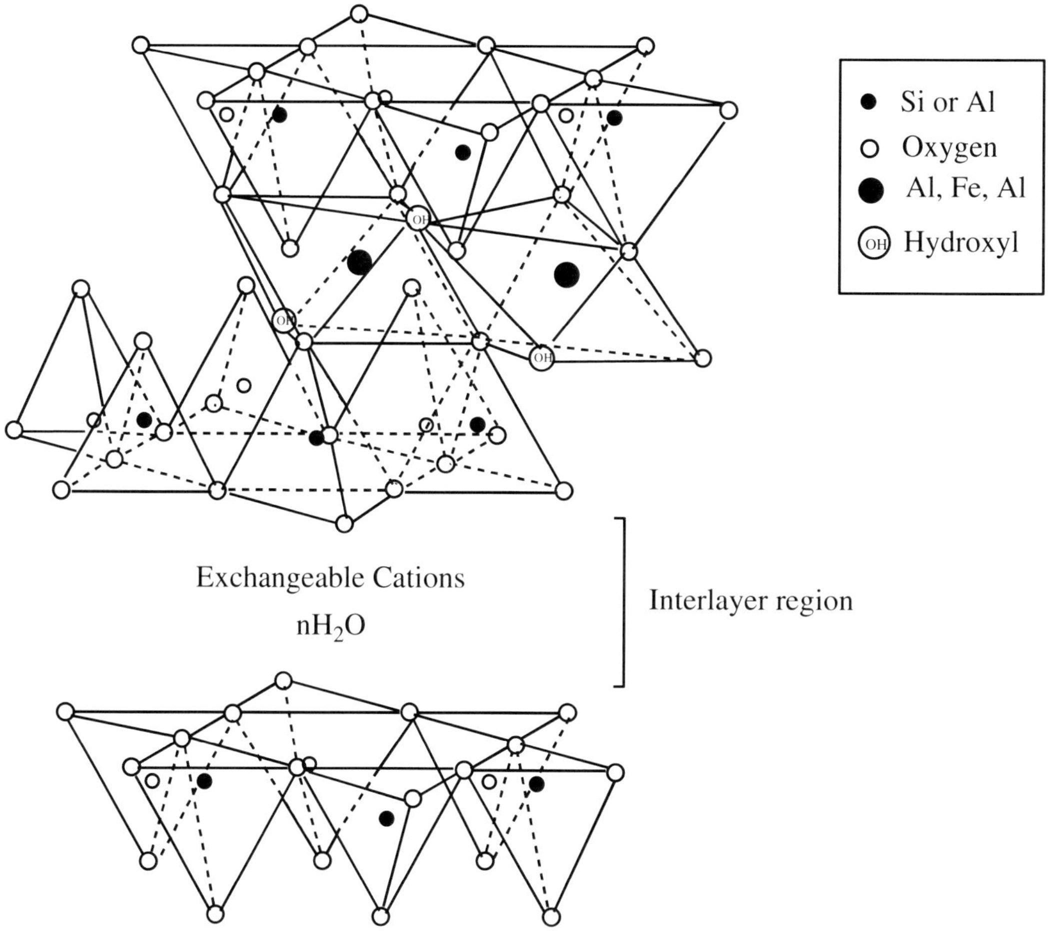

Figure 3. Structure of montmorillonite.

Types of Bentonite

European Bentonite Association (EUBA) has classified bentonites into following types:

Sodium Bentonite (Swelling-type)

Sodium-bentonite is a naturally occurring rock in which sodium ions occupies the intermediate layers predominantly (Figure 4). It can absorb up to 5 times of its original weight in water and swell up to 12-15 times its dry volume due to which it is also known as swelling type bentonite [19]. Calcium or Magnesium ions can also occur commonly in sodium-bentonite in varying concentrations. The property of swelling makes sodium bentonite useful as a sealant, since it provides a self-sealing, low permeability barrier. It is used in drilling mud for oil and gas wells and boreholes for geotechnical and environmental investigations due to its excellent colloidal properties [20]. Various surface modifications to sodium bentonite improve some rheological or sealing performance in geoenviromental applications[20].

Calcium Bentonite (Non-Swelling-Type or Calcium)

Calcium-bentonite or non-activated bentonite is also a naturally occurring rock, in which the intermediate layers are occupied by Ca^{2+} or Mg^{2+}-ions predominantly (Figure 5). Calcium bentonites are also dispersed in water but particles are coarser than those of the swelling bentonites [21]. It is the main active ingredient of fuller's earth, which is one of the earliest industrial cleaning agents [22]. Calcium bentonite may be converted to sodium bentonite to exhibit many of sodium bentonite's properties by ion exchange. The viscosity and fluid loss of suspensions, of sodium-beneficiated calcium bentonite may not be fully equivalent to those of natural sodium bentonite due to residual calcium carbonates. Calcium bentonite is used as bonding agents in foundry molding sand, mineral binders and to decolorize oils.

Acid Activated Bentonite

Acid activated bentonite is obtained by partially dissolving bentonite (typically the Ca^{2+} form) using acidic solutions to replace Ca^{2+} with protons and to leach out Al^{3+}, Fe^{3+}, and Mg^{2+} from the silicate layers (Figure 6) [23]. It exhibits high surface area, porosity, and surface acidity. Protons from the acid replace the exchangeable cations first, and then attack the layers. In contrast to smectites saturated with metal cations, these proton-saturated smectites (H-smectites) are unstable. The strongest acid sites are connected with free protons present in the dispersion, while the weaker acid sites are associated with the cations released from the structure [24]. Acid-activated bentonites are relatively inexpensive solid sources of protons. Acid activated bentonites has been used to adsorb impurities from edible and industrial oils, catalysts, carbonless copying paper and catalyst supports [25].

Organophilic Bentonite (Organoclays)

Organophilic bentonite or organoclays are the modifications of natural bentonite. In organo-bentonites the cations in the intermediate layers are replaced by polar organic molecules (e.g., alkylammonium ions, Figure 7). They are prepared by the reaction of bentonite with various aliphatic ammonium salts. They are hydrophobic and can swell in organic solvents. The degree of solvation depends on the extent of the surface coating of the clay particles by organic matter; the degree of saturation of the exchange capacity of the clay mineral by organic cations; and the nature of the organic liquid [26].

The organophilic feature reduces the energy of the clay mineral and makes it more compatible with the organic polymers. The addition of organoclays into polymeric matrices improves mechanical, physical (thermal and barrier) and chemical properties of the matrices and in some cases reduces cost [28]. Organoclays also have been used as adsorbents, rheological control agents, paints, grease, cosmetics, personal care products, oil well drilling fluids, etc. [29, 30].

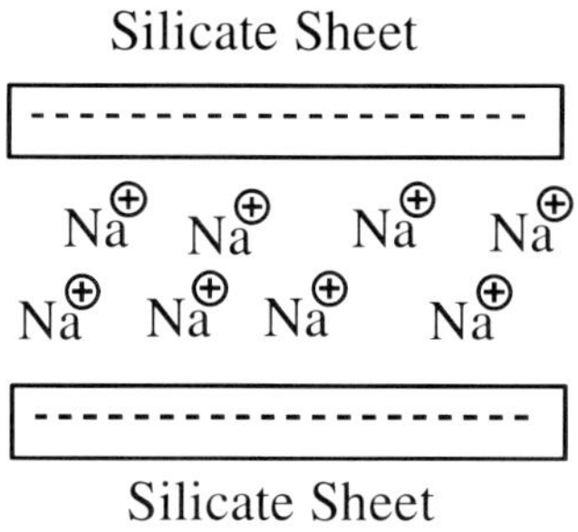

Figure 4. Sodium bentonite.

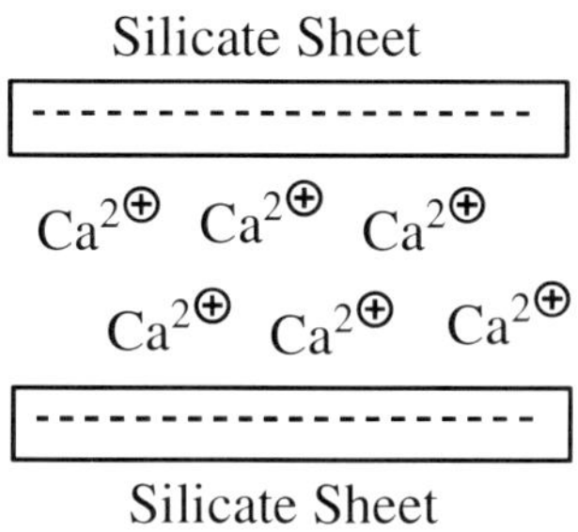

Figure 5. Calcium bentonite.

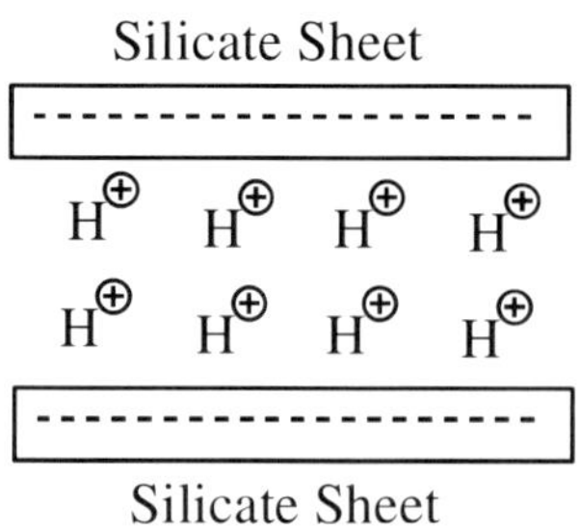

Figure 6. Acid activated bentonite.

General Properties of Bentonites

Fresh bentonite is white to pale green or blue and, with exposure, darkens in time to yellow, red, or brown. Specific gravity of dry bentonites is in between 2.4 and 2.8. The chemical composition of bentonite shows considerable variations (SiO_2 is found to vary in range 45-65%, Al_2O_3 17-25%, Fe_2O_3 up to 12%). Physical characteristics of bentonite are affected by nature of water layer in the montmorillonite. Loss of absorbed water between the silicate sheets takes place at relatively low temperatures (100-200 °C). Loss of structural water (i.e., the hydroxyls) begins at 450-500 °C and is complete at 600-750 °C. Further heating to 800-900 °C disintegrates the crystal lattice and produces a variety of phases, such as mullite, cristobalite and cordierite, depending on initial composition and structure. The ability to take up water affects the utilization and commercial value of bentonite.

Interaction with Water

When montmorillonite comes in contact with water, molecules of water penetrate between the interlayer spaces leading to the swelling. A stable ordered configuration may be retained in montmorillonite layers, when up to four monomolecular layers of water fill the space between the platelets. The volume of the hydrated clay may be up to 12-15 times larger than its dry bulk volume [31, 32]. The ability of montmorillonites to swell upon interaction with water determines its rheological and ion exchange properties of these minerals.

Rheological Properties

In aqueous suspensions, montmorillonite platelets are evenly dispersed forming a viscous colloidal solution. Platelets have a total negative charge on their faces and positive charges on their edges (FE type). Platelets orient themselves in the negative-to-positive conformations and form a jelly like structure. The inter-platelet space of the bentonite is filled by the electrolyte solution of cations that gives rise to a variety of useful property to its suspension such as swelling ability, high viscosity and high suspending power, thixotropic property with a good gelling strength. Surfactants have been used in sodium bentonite to improve the rheological properties [33, 34].

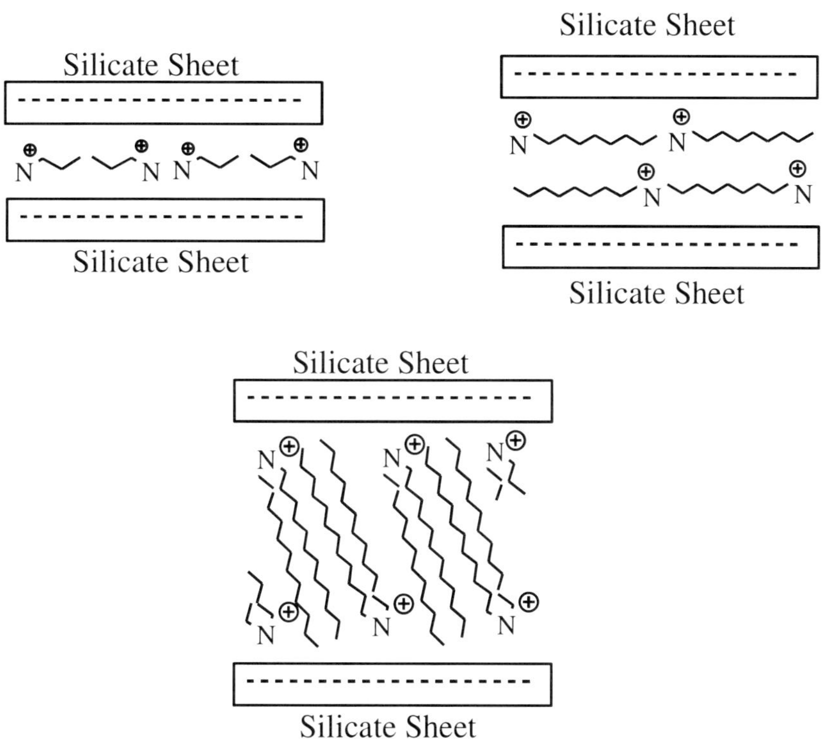

Figure 7. Orientations of alkylammonium ions in the galleries of layered silicates: (a) monolayer, (b) bilayers, (c) paraffin-type arrangements of alkylammonium ions with different tilting angles of the alkyl chains [27].

Sorptive Properties

Bentonite clays have a high adsorption and absorption property. When calcined, all clays have both a large pore volume and a high surface area. Surface areas of large external and internal "channel" of bentonite are the highest of any naturally-occurring minerals known. Temperature, pH, charge per unit area of platelets and a composition of ions in the interlayer space affect sorption processes in bentonites.

Surface Area

Bentonite possesses large surface area due to which it can act as catalysts, or as a carrier for catalysts. For example, an aluminum and silicon catalyst used in petroleum refining is produced from clay by treatment with sulfuric or hydrochloric acids to remove adsorbed ions like alkaline and alkaline-earth metals, and iron, as well as some of the magnesium and aluminum from the smectite lattice. The specific surface area of finely milled bentonites can be as large as 300 m^2/g.

Ion Exchange

In bentonites the substitutions of silicon by cations produce an excess of negative charges in the lattice, which is balanced by cations (Na^+, K^+, Mg^{2+}, Ca^{2+}) in the interlayer space. These cations are exchangeable due to their loose binding and, together with broken bonds (approximately 20% of exchange capacity) give montmorillonite a high cation exchange capacity, which is little affected by particle size. This cation exchange capacity allows the mineral to bind not only inorganic cations such as caesium but also organic cations such as the herbicides diquat, paraquat [35] and triazines[36] and even bio-organic particles such as rheoviruses [37] and proteins [38] which appear to act as cations. Variation in exchangeable cations affects the maximum amount of water uptake and swelling. The interlayer spacing is reduced when monovalent sodium ion is replaced by divalent calcium ions. This can be attributed to the fact that on ion exchange the hydrated double layer of sodium ion is replaced by the less hydrated layer of calcium ions. Moreover sodium ion has twelve water molecules in the hydration shell while calcium ion has only eight water molecules of hydration (Figure 8).

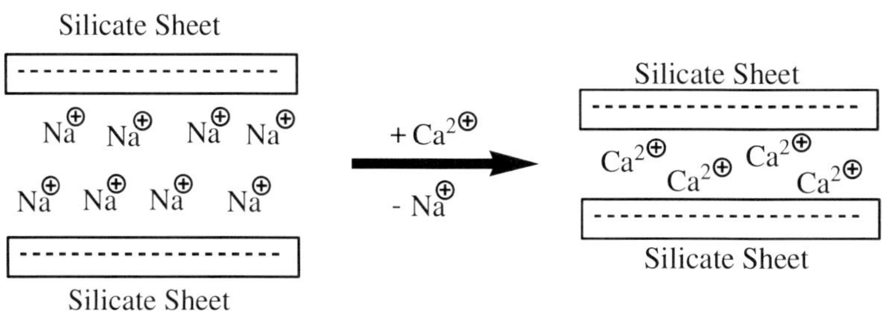

Figure 8. Reduction in interlayer spacing from Sodium bentonite to calcium bentonite.

Exploration of Bentonite

Bentonite occurs on every continent except Antarctica, however, depending on its properties and types of accessory minerals present, can vary widely from deposit to deposit, even within the same geographical location. Most commercially significant bentonite deposits were formed by the devitrification of volcanic ash that fell into shallow oceans or brackish ponds and lakes [39, 40]. These deposits often cover large areas and may consist of several layers, representing multiple volcanic eruptions. Some bentonite deposits are formed by hydrothermal groundwater alteration of felsic host rocks. Many bentonite deposits are identified where they have been exposed at the ground surface by erosion. The layers, or beds, of bentonite have unique surface texture, which results from swelling and shrinking of the clay as it undergoes repeated cycles of wetting and drying. Ca-bentonite deposits posseses an alligator hide appearance, while Na-bentonite deposits are frequently characterized by have pronounced popcorn texture. Generally commercially viable bentonite deposits are typically 0.5 to 5 meters thick but may exceed 30 meters in some locations. A detailed knowledge of the physical and chemical characteristics of the clay in each deposit is important to meet the diverse quality criteria for each of the many end uses for bentonite. These physical and chemical data are gathered from pre-mine drilling and testing. In this way, the different qualities of bentonite within a bed are identified, selectively mined, and stockpiled. Exploration data, together with information on soils, vegetation, wildlife, local land use, and archeological, cultural, and groundwater resources, are used to develop detailed mining and reclamation plans.

Mining of Bentonite

Bentonite is usually extracted by surface mining. Bulldozers and tractor-scrapers are used to remove the overburden material; the topsoil and subsoil are stockpiled separately so they can be replaced during reclamation. The remaining overburden above the clay is then removed and stockpiled (Figure 9).

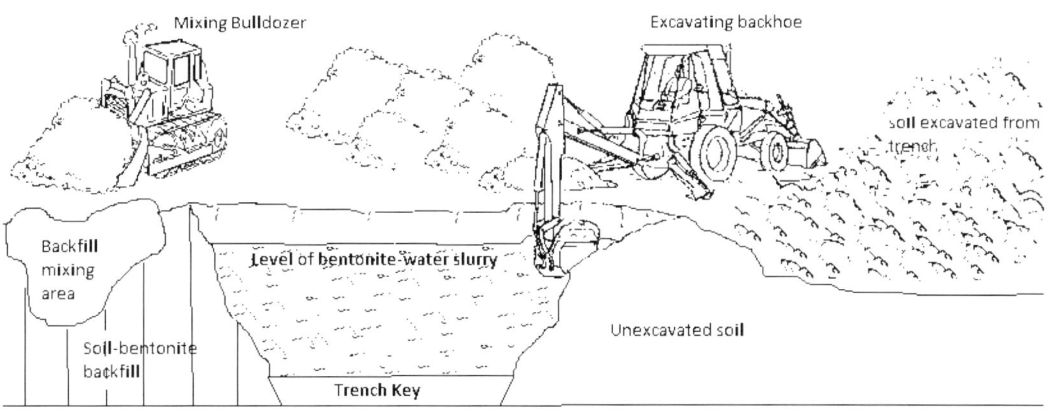

Figure 9. Mining of bentonite.

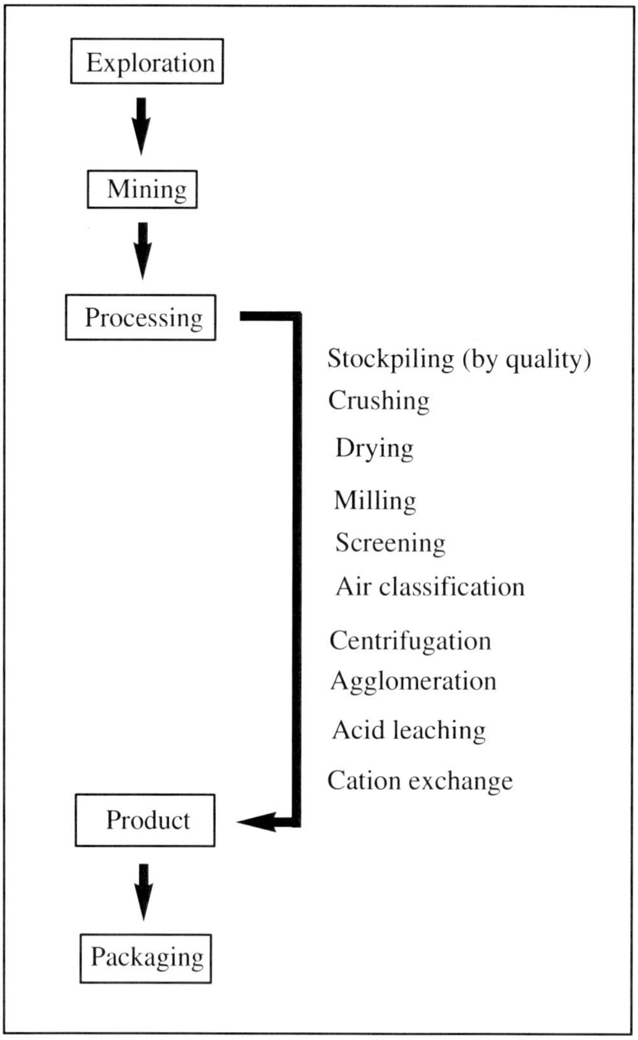

Scheme 1. Steps involved in the processing of bentonite.

After the clay has been extracted from the pit, the overburden, subsoil, and topsoil are replaced, and the surface is contoured to blend with the surrounding terrain and reestablish drainage through the area. In cast-back mining, materials removed from one pit are placed directly into an adjacent pit to minimize the disturbed area that also minimizes the need for stockpiles and shortens the time required for reclamation. With this technique, it is not uncommon for a pit to be opened, closed, and seeded within a year.

BENTONITE PROCESSING

After the removal from the mine, bentonite is taken to a processing plant and stockpiled by quality. At the plant, clays of different qualities are frequently blended during processing in order to achieve more consistent finished product quality. As per the requirement in the final application, bentonite is either sieved (granular form) or milled (into powder and superfine powder form).

Common bentonite processing steps include extrusion, drying, milling, screening, air classification, centrifugation, agglomeration, acid leaching, and cation exchange (Scheme 1).

Bentonite is purified by removing the associated gangue minerals, or treated with acids to produce acid activated bentonite (bleaching earths), or treated with organics to produce organoclays. Processing is designed to maximize the dispersibility of the clay, increase its surface area and alter its surface chemical properties. The surface area of bentonite is greatest when the individual montmorillonite sheets are fully delaminated and dispersed. Mechanical shear and cation exchange are used to delaminate montmorillonite. The water dispersibility and colloidal stability of the clay are improved when exchangeable divalent surface cations, such as Ca^{2+} and Mg^{2+}, are replaced by univalent cations, e.g., Na^{+} and Li^{+} [41]. Where dispersion of clay in nonpolar fluids is required, hydrophilic Na-bentonites are transformed to hydrophobic organobentonites by exchanging the surface cations with positively charged organic species, most commonly quaternary amines. Organobentonites are widely used in the development of clay-polymer nanocomposites [42]. In addition, various oxidation, reduction, and magnetic separation processes have been developed to improve purity.

Applications of Bentonite

Bentonite has a broad range of applications. Major uses include binding foundry sand, absorbing grease, oil, and animal wastes, pelletizing iron ore, and improving the properties of many drilling muds. It is also used as an ingredient for ceramics; waterproofing and sealing in civil engineering projects, serving as a filler, stabilizer, or extender in adhesives, paints, cosmetics, medicines, and other products, as a carrier in pesticides and fertilizers, and as a bonding agent in animal feeds; clarifying wine and vegetable oil and purifying wastewater. Small amounts of bentonite are also used as a catalyst in the refining of petroleum (Figure 10).

Metal Casting

Metal casting is among the most important current uses of bentonite. Approximately 25% of all worldwide production is used by the metal casting industry. More than 70% of iron castings are made in moulding sands with bentonite. After filling a mould with liquid cast iron a part of montmorillonite contained in bentonite undergoes a thermal degradation, loosing its binding properties (bentonite deactivation process), forming chamotte (Figure 11). The moulding sand separated from casting is subjected to rebounding and once more introduced into the production process. Rebounding is done by addition of fresh high-silica sand and bentonite [43]. Engine blocks, brake drums, transmission housings, manhole covers, street drains, brass water fittings, cast-iron pans, and more are produced from foundry molds containing bentonite. Bentonite is used as a bonding agent in green sand molding process in most ferrous and many nonferrous foundries. Both Na- and Ca-bentonites are used in this process, either individually or as blends. The type and amount of bentonite used depends on the metal being cast, the size and complexity of the casting, and the type of molding equipment.

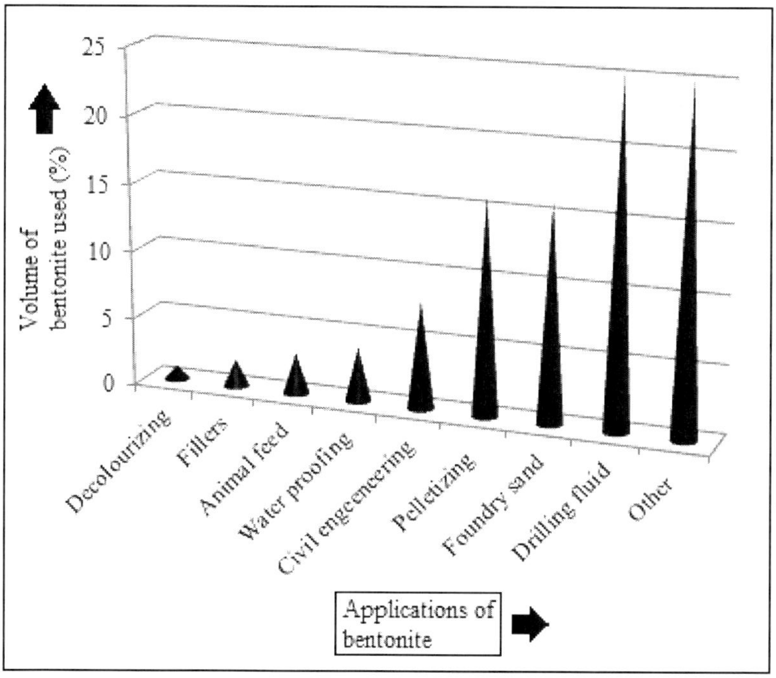

Figure 10. Plot of applications of bentonite *verses* volume of bentonite used (in percentage).

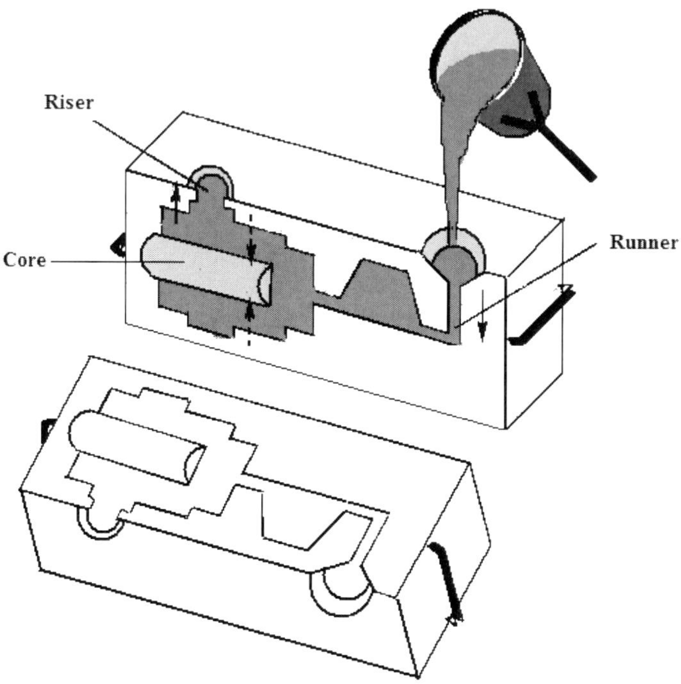

Figure 11. Bentonite in metal casting.

In comparison with other binding materials, the adhesive characteristics of bentonite provide a superior bond for sand in the mold prior to, during, and after the molten metal is poured, making metal casting rapid, highly flexible, and cost effective [44].

Iron Ore Pelletizing

Bentonite is the most common binder for iron ore pelletizing. It holds the pellet together by forming a solid bridge of hardened gel, strengthening the particle contact points and increasing the van der Waals forces between the particles [45]. Bentonite is used primarily for improving both plasticity and the compressive strength of iron ore pellets so that they will not break during handling on the traveling grate from the balling drum to the kiln, drying, and firing in the kiln [46 ,47, 48]. Usually, 5-7 kg per ton of the wet magnetite concentrate of bentonite is used for pelletizing. They can be homogeneously dispersed in the concentrate and posseses high moisture absorbing capacity in the balling feed therefore they are used in iron ore pelletizing processes. The pellets also allow gas flow through reduction furnaces in the steel-making process. In some cases, iron ore concentrates mixed with bentonite are compacted under high pressure to make briquettes.

Pet-Waste Absorbents

Domestic cats are among the most popular pets in the U.S. As the majority of these cats live primarily indoors, litter boxes containing some type of absorbent litter material are needed [49]. Sodium bentonite clay is able to absorb more water than bentonite clay containing other ions, such as calcium therefore it is used for making clumping cat litter (Figure 12). It encloses pet wastes in clumps of absorbent clay, which can be scooped easily and completely from the litter box, decreasing the amount of litter wasted and keeping litter boxes clean and odor-free [50].

Figure 12. Bentonite as cat litter.

DRILLING FLUIDS

Drilling fluids are primarily water-bentonite suspensions. They are used in oil, gas and geothermal drilling industry because they perform many functions like transporting rock cuttings to surface, lubricating the drill bit, applying hydrostatic pressure in the well bore to ensure well safety [51]. Also, it is used in horizontal directional drilling to place utility cables and pipelines under buildings, streets, highways, and environmentally sensitive areas without the need for costly and disruptive surface trenching. The bentonite in the fluid also functions to limit loss of the fluid into the soil and rock around the drill hole. Organoclays has also been used in non-water-based drilling and in many emulsion-based fluids [52].

BLEACHING AND CLARIFYING

Acid activated Ca-bentonites are used to clarify and decolorize, animal, vegetable, and mineral oils that contain many impurities that can produce undesirable oxidation products and affects product shelf life [53, 54]. Certain oils are clarified by allowing percolation through beds of bentonite in which a selective adsorption of organic compounds onto the clay mineral surfaces. The portions of the oil which are darkest in color are removed preferentially by the clay [55]. Na-bentonite is also used as a fining agent for clarifying wine and juice to remove suspended impurities. Because of their small size and solubility, these undesirable materials are difficult to remove by conventional filtration [28]. The large surface area and high negative charge of montmorillonite makes bentonite ideal for adsorbing positively charged compounds, such as proteins, and removing them through flocculation.

DESICCANT

Bentonites absorb moisture from air and therefore can also be used as a desiccant [56]. Desiccant-grade bentonites contain Ca^{2+} and Mg^{2+} as the dominant exchangeable cations and always have high montmorillonite contents with high layer charge (Figure 13).

Figure 13. Bentonites as dessicants.

Although Ca-Mg-bentonites have little swelling capability in water, they function well as desiccants because of the greater first and second shell hydration energies of Ca^{2+} and Mg^{2+} cations relative to Na^{+}. They are used to protect pharmaceutical, nutraceutical and diagnostic products from moisture degradation and extend shelf life. They offer a higher adsorption capacity than silica gel desiccants.

Paper Making

Bentonite is widely used in combination with cationic polymers to flocculate cellulose fibers during paper making. Its use improves several key processes on the wet end of paper machines, including solids retention, drainage rate, and uniform sheet formation [57]. The addition of only 0.2 weight percentage bentonite can increase paper production by as much as 30% in a paper mill. Acid-activated bentonites are used in making carbonless copy paper, which is composed of three sheets [58]. The upper surfaces of the middle and lower sheets are coated with clay, which serves as a solid acid that reacts with microencapsulated dyes fixed on the undersides of the upper two sheets. When pressure is applied to the top sheet, the microcapsules of colorless dye break and react with the acidic clay to form a permanent colored dye.

Environmental Sealants

Bentonite absorbs water and swells many times of its dry volume, and extremely low water permeability makes it an excellent sealant for environmental applications. Swelling ability along with very small particle size, allow bentonite to plug even the smallest voids in the soil, creating a barrier of very low permeability. A variety of granular and powdered bentonite products have been developed for sealing landfills, sewage lagoons, and decorative ponds. Bentonite chips can rapidly sink through standing water and, once in place, they swell to apply significant pressure against the cavity walls, forming a secure seal. It is also used to seal pores in dams and other constructions. Pumpable, high-solids bentonite grouts may be used where free space is limited, which are prepared from bentonite that has been specially modified to delay onset of gelling when they are mixed with water, allowing them to be emplaced while still in a fluid state. Bentonite grouts may also be injected under high pressure directly into soil to fill pores and form a low-permeability seal for building foundations and other subterranean structures [59].

Civil Engineering

Na-bentonite has been used to create low-permeability *in situ* barriers to groundwater movement and to form *in situ* foundations for buildings. At bulk densities between 1.1 and 1.3 g/cm^3, the hydraulic conductivity of smectite gels is sufficiently low to make them useful as effective grouts for sealing rock fractures. Bentonite-based products are also used in boring tunnels for highways and subways, and for installing pipelines to transport drinking water, storm water, and sewage. In these applications, bentonite is used in tunnel-boring machines to stabilize the excavation face, lubricate the machine as it moves through the earth, and for sealing the tail of the machine to prevent flooding where boring is performed below groundwater level.

ADHESIVES THICKENER

A bonding agent must have certain adhesive properties, and these are developed when bentonite is mixed with only a small amount of water. When bentonite is mixed with water and is added as a constituent to a combination of materials, rigidity rather than plasticity is developed in the final product. Some liquid adhesives are strong, but are of low viscosity due to which its efficiency is reduced as they soak too quickly into porous papers. When sodium bentonite is added to a less viscous adhesive instantly a thickened adhesive is formed and hence the penetration into the paper is controlled. Bentonite or hectorite is usually just blended with a liquid adhesive to improve its gelling and increase the consistency, making spreading and removal easier [60].

EMULSION STABILIZER

Due to the large surface area of bentonites impart stabilization of aerosols and cleaning compounds. Bentonite is used as emulsion stabilizer. Bentonite is compatible with organic gums and associative thickeners modify the consistency and feel of organic thickeners by reducing tackiness, gumminess and stringiness. Emulsions are formed by mixing and agitation of two immiscible liquids, such as oil and water. One liquid becomes a continuous phase, and the other liquid forms as droplets or globules within it. The ratio in which the two liquids are mixed determines the role of each in the emulsion. When the oil-water interface is strengthened, the stability of the emulsion is increased. Bentonite has a stabilizing effect on emulsions when it is added to the mixture prior to agitation. Apparently, the clay flakes form a thin film at the oil-water interface, and the large surface area of the clay aids in the entrapment of the oily globules as they are broken down by mechanical agitation. Detergents consist of an alkali to loosen dirt and grease, an emulsifier to hold oil and grease, and a protective colloid to prevent soil particles from redeposition in the fabric. Sodium bentonite provides all of these actions, although it is very mild in alkalinity but active as an active agent when added to detergent formulas.

CATALYSTS

The surface acidity, plays a very important role in the application areas of bentonites, especially when they are used as catalysts. It may be adjusted to the required value by applying either controlled acid activation or controlled thermal treatment [61, 62]. Bentonite is stable up to only 600°C and posses Bronsted acidity. It is a good catalyst for benzaldehyde acetalization glycerol hydrogenolysis [63] and catalytic degradation of municipal waste plastics to produce fuel range hydrocarbon [64]. Certain modified bentonites have been used as catalysts such as KOH-bentonite for *trans*-esterification [65], bentonite based Fe-nanocomposite (Fe-B) for Photo-Fenton discoloration, mineralization of an azo-dye Orange II [66] and bentonite/PS-SO_3H nanocomposites for the Biginelli reaction [67]. Recently the pellet-type Zr-SBA-15/bentonite has been used in the production of bio-diesel from waste cooking oil on a packed bed reactor [68]. Modified bentonite with ionic liquid showed higher catalytic activity than sulfuric acid and catalyzed the esterification of oleic acid to its ethyl ester [69].

Cation Exchanger

Sodium smectite clay, serves as a very effective cation exchanger. Bentonite and zeolites have large cation exchange capacities, which enable them to be applied as natural sorption materials for organic and ionic contaminants. Bentonite treatment has a substantial effect on the metal uptake [70]. The commercial natural bentonite has been used to remove silver ions from aqueous solution at optimum pH, nickel from waste water, the adsorption of nickel increases with an increase in solution pH from 2.0 to 9.0[71]. Bentonite also adsorbs Pb(II) ions which changes in unit cell dimensions and symmetry of the parent bentonite [72]. The high cation exchange capacity enhances the retention of wastes, especially heavy metals. In addition, a mixture of sodium bentonite and soil forms tough, flexible mastic that is highly durable and not easily ruptured. Bentonites provides an adequate surface area for the adsorption of the insecticide, leading to even distribution of the insecticide when scattered over a large area, increased durability or life of the insecticide, and also increased effectiveness by keeping the insecticide on the vegetation rather than on the ground. Sodium bentonites, soak easily with poisons dissolved in an organic solvent. This toxicant solution is sprayed on Volclay and mixed until the solvent has evaporated. This leaves a film of poison soaked in and on the Volclay products. Powder insecticides may be applied by airplane or ground applicators. They have the ability to penetrate dense foliage and are useful in corn borer and mosquito control.

Food Additives and Preservatives

Bentonites can increase pellet durability when added at concentrations between 1-2% complete feed. Bentonite fining can lead to sluggish fermentation in some juices if the bentonite lees are removed [73]. It is used to clarify apple juice by the adsorption on the surface [74] and as anti-caking agents in food. It is added to feed contaminated by radioactive fallout or made available to grazing animals will reduce levels of radiocaesium in animals and their products. Poultry feed supplemented with bentonite can improve growth performance and nutrient digestibility in broilers chicks [75, 76]. Mangos coated with bentonite or bentonite loaded with potassium sorbate at ambient temperature could extend the shelf-life of mangos with optimum quality [77].

Nuclear Waste Disposal

Bentonite-sand mixtures are used as bentonite-based backfill materials for high-level radioactive waste repositories. They are also used in Japan as artificial barriers in disposal facilities for low-level radioactive waste [78]. Its low hydraulic conductivity, micro-porous structure, good sorption properties and swelling capacity make this as an effective barrier-protecting the canister and restricting the movement of radionuclide released from the waste packages after canister failure [79].The main function of their materials is to create a zone of very low hydraulic conductivity around wastes and to separate wastes from the surrounding environment. Bentonite strongly adsorbs Sr, Cs and Am from solution. Sorption by basalt/bentonite mixtures is strongly dependent on redox state, with large sorption capacities reported for U, Np and Tc under reducing conditions. Desorption of U, Np, Se and Tc is much less efficient than initial sorption. The sorption of radionuclides by colloidal bentonite may have a significant effect on waste package performance [80].

Medicinal Application

Bentonites have been used in traditional medicines [81]. The vast surface area and chemical variability of hydrothermally altered bentonites may provide a natural pharmacy of antibacterial agents. It is reported to be one of the most powerful healing clays. Clay has been used as bandaids to deliberate consumption that reduces borborygmi and intestinal ailments, anecdotal accounts of therapeutic clays. It has been prescribed as a bulk laxative, and it is also used as a base for many dermatologic formulas [82]. Granular bentonite is being studied for use in battlefield wound dressings [83]. Some bentonites offer distinct antibacterial properties. Copper incorporated bentonites are shown to produce bactericidal properties on the *Escherichia coli* [84]. Antibacterial effects might result from physical interaction and/or chemical interaction of the clay with bacteria. It has been used in detoxifying victims of paraquat poisoning and to reduce the transfer of radiocaesium to milk and other animal-derived foods [85]. It has been used for the treatment of rheumatic and orthopedic disorders [86], and as a digestive aid worldwide. It has been proven to have a very strong effect against aflatoxins. These are naturally occurring mucotoxins that are produced by many species of the fungus like *Aspergillus*. The addition of bentonite to aflatoxin-contaminated diet diminished most of the deleterious effects of the aflatoxins [87].

Miscellaneous Applications

Bentonite suspensions are also used as fire retardant material, particularly in combating forest fires. When water stream is directly implemented on the vegetation, such as a tree, much of the water ricochets off and is lost owing to its high fluidity. But on addition of bentonite to water its viscosity increases and a greater proportion of liquid is retained on the vegetation. Because of the thicker consistency, suspensions can be dropped from planes to aid fire-fighting. Thus a greater area can be covered in a shorter period of time. Bentonite is also of cosmetic products, such as paste masks, skin care and cleansing preparations, eyeliners, foundations [88]. Greases utilising bentonite are generally regarded as high temperature greases and are used in certain vehicle wheel bearings where disc brakes are incorporated. These bentonite bases have no "Drop Point" and this makes them eminently suitable or this application. The organoclays are used as gelling agents in paints, lubricants, and other oil-based media. The modified bentonites has been used for the removal of dyes such as methyl orange [89], congo red [90] from the wastewater.

FUTURE TRENDS

Demand for bentonite in different applications is increasing day by day therefore its production will have a huge impact on the economic growth of the countries. As the demand is increasing, the mining and processing technologies will also develop. In mining and processing more efficient methods are expected to develop that require less time with greater selectivity and low cost with a high level of quality control of the clay. The unique properties of bentonite are expected to be used to develop newer materials for different purposes such as energy production, human health, nanotechnology etc.

The past decade has seen a significant rise in research on clay minerals and their role in animal and human health [91].The use of bentonite in the medical applications is also expected to increase. Clay minerals have long been reported to have healing properties, but their true efficacy and the mechanisms by which they work is yet to be investigated further. In comparison with antibiotics, inorganic antimicrobial minerals are considerably more stable and heat resistant. Identification of the combinatorial chemistry of solutions buffered by natural, potentially bioactive mineral resources could result in the discovery of new antibacterial agents which can fight with the existing antibiotic-resistant infections and diseases. The new focus on medical mineralogy and bentonites, will progress because novel *in vitro* and *in vivo* experiments with clay minerals have much to offer ahead. Research on nanoscale clay-plastic composites is expected to accelerate by nanoscale interactions among materials. Major areas of development include high-performance sporting equipment, low-permeability food-packaging films, and lightweight automobile parts. The use of bentonite as a functional ingredient in cosmetics, as a carrier for pesticides to improve efficacy and reduce usage, in insulating compounds for electrical components, in electrically conductive compounds for use in grounding systems and for removing contaminants from wastewater is also expected to increase.

CONCLUSION

Bentonites are colloidal clays composed of clay mineral montmorillonite, resulting from the decomposition of volcanic ash. They are highly water absorbent with high plasticity, and have high swelling and shrinkage characteristics. The surface properties can be highly modified by simple ion exchange mechanisms. Bentonites are used in papermaking technology, cut-off walls, wastewater purification and in different aspects of nano-technology. Modified clays are used in other applications such as adsorbents of organic pollutants in soil, water and air, rheological control agents, paints, cosmetics, refractory varnish, thixotropic fluids, etc. They are also used as feed and food additives. Bentonite can be used in medical applications such as a gel, bulk laxative and also as a base in many dermatological formulations. Copper incorporated bentonites showed bactericidal properties on *E. coli*. Bentonites have been found to reduce levels of radioactive elements in animals and their products. It is effectively used as backfill materials for high-level radioactive waste disposal.

REFERENCES

[1] Parker, S. P. (1988). McGraw-Hill Encyclopedia of the Geological Sciences, 2nd ed. New York, McGraw-Hill.
[2] Parkes, W. R. (1982). Occupational lung disorders, London, Butterworths.
[3] Khan, A. D., Ijaz, N. & Shahzad, K. (2001). *Agro Vet. News*, 2.
[4] Hassan, M. S. & Abdel-Khalek, N. A. (1998). *Appl. Clay Sci.*, 13, 99.
[5] Khan, A. D., Ahmad, R., Shahzad, K. & Saeed, S. (2004). Proc. Int. Nut. Conf., 1-2 December, (Lahore-Pakistan) 127.
[6] Saidi, A., Shamanian, M., Barati, M. & Azari, K. (2004). *Internat. J. ISSI*, 1, 38.
[7] Bekkour, K., Ern, H. & Scrivener, O. (2001). *Appl. Rheology*, *11*, 178.
[8] Diaz, F. V. & Souza-Santos, P. (2001). *Quimica Nova*, *24*, 345.
[9] Torok, A. & Thompson, T. D. (1972). *Transactions SME-AIME*, *252*, 15.
[10] Venkatathri, N. (2006). *Bull. Cat. Soc. India*, *5*, 61.
[11] Pusch, R. (1992). *Clay Minerals*, *27*, 353.
[12] Cara, S., Carcangiu, G., Padalino, G., Palomba, M. & Tamanini, M. (2000). *Appl. Clay Sci.*, *16*, 125.
[13] Cara, S., Carcangiu, G., Padalino, G., Palomba, M. & Tamanini, M. (2000) *Appl. Clay Sci.*, *16*, 117.
[14] EFSA Panel on Additives and products or substances used in animal feed (FEEDAP), *EFSA Journal*, *10*, *2787*, 2010.
[15] Mine bureu, 18. <ibm.gov.in/imyb2010_bentonite.pdf>
[16] Vimonses, V., Jin, B., Chow, C. W. K. & Saint, C. (2009). *J. Hazard. Mat.*, *171*, 941.
[17] Vimonses, S., Lei, B., Jin, C. W. K., Chow, & Saint, C. (2009). *Chem. Eng. J.*, *148*, 354.
[18] Ozcan, A. S., Erdem, B. & Ozcan, A. (2005). *Coll. Sur. A: Physico. Eng. Asp.*, *266*, 73.
[19] Lim, S. C., Gomes, C. & Kadir, M. Z. A. (2013). *Int. J. Electrochem. Sci.*, *8*, 11429.
[20] Luckham, P. F. & Rossi, S. (1999). *Adv. Coll. Inter. Sci.*, *82*, 43.
[21] Johnstone, S. & Johnstone, M. (1961). Minerals for the chemical and allied industries. Chapman and Hall. 2nd ed., London.
[22] Vimonses, V., Jin, B., Chow, C. W. K. & Saint, C. (2010). *Chem. Eng. J.*, *158*, 535.
[23] Kotai, L., Sajo, I. E., Jakab, E., Keresztury, G., Pfeifer, E., Kocsis, L., Papp, K., Gacs, I. Valyon, J. & Lippart, J. (2008). *Chem. Lett.*, *37*, 1110.
[24] Komadel, P. & Madejova, J. (2006). *Dev. Clay Sci.*, *1*, 265.
[25] Fernandes, C., Catrinescu, C., Castilho, P., Russo, P. A., Carrott, M. R. & Breen, C. (2007). *Appl. Cat. A: General*, *318*, 108.
[26] Kakegawa, N. & Ogawa, M. (2002). *Appl. Clay Sci.*, *22*, 137.

[27] Bergaya, F., Theng, B. K. G. & Lagaly, G. (2006). Handbook of Clay Science, Ist ed. Elsevier.
[28] Markarian, J. (2005). *Plast. Add. Compound*, 18-25 November.
[29] Beall, G. W. & Goss, M. (2004). *Appl. Clay Sci.*, *27*, 179.
[30] Araujo, S. S., Andrade, D. L. A. C. S., Carvalho, L. H., Gomes, A. S. & Silva, S. M. L. (2005). *Plastico Ind.*, 104.
[31] Harben, P. W. & Kuzvart, M. (1996). Industrial minerals: A global geology, Industrial Minerals Information Ltd, Metal Bulletin, London.
[32] Meunier, A. (2005). Clays. Springer, Berlin, Heidelberg, New York.
[33] Gungor, N. & Ece, O. I. (1999). *Mat. Lett.*, *39*, 1,.
[34] Gunister, E., Isci, S., Oztekin, N., Erim, F. B., Ece, O. I. & Gungor, N. (2006). *J. Coll. Inter. Sci.*, *303*, 137.
[35] Weber, J. B., Perry, P. W. & Upchurch, R. P. (1965). *Soil Sci. Soc. Am. Proc.*, *29*, 678.
[36] Weber, J. B. (1970). *Soil Sci. Soc. Am. Proc.*, *34*, 401.
[37] Lipson, S. M. & Stotzky, G. (1983). *Appl. Env. Micro.*, *46*, 673.
[38] Potter, E. V. & Stollerman, G. H. (1961). *J. Immunology*, *87*, 110.
[39] Christidis, G. E. & Huff, W. D. (2009). *Elements*, *5*, 93.
[40] Elzea, J. M. & Murray, H. H. (1990). *Appl. Clay Sci.*, *5*, 229.
[41] Murray, H. H. (1991). *Appl. Clay Sci., 5*, 379.
[42] Carrado, K. A. & Komadel, P. (2009). *Elements*, *5*, 111.
[43] Holtzer, M., Grabawoska, B., Bobrowski, A., Zymankowaska-Kuman, S. (2009). *Arch. Foundry Eng.*, *9*, 69.
[44] Sanders, C. A. (1973). Foundry Sand Practice. 6th ed. American Colloid Company, Skokie, Illinois.
[45] Hayes, P. C. (2003). Process principles in minerals and materials production. Hayes Publishing CO, Queensland, Australia.
[46] Eisele, T. C. & Kawatra, S. K. (2003). *Miner. Process. Extr. Metall. Rev.*, *24*, 1.
[47] Forsmo, S. P. E., Apelqvist, A. J., Bjorkman, B. M. T. & Samskog, P. O. (2006). *Powder Tech.*, *169*, 147.
[48] Kawatra, S. K. & Ripke, S. J. (2001). *Min. Eng.*, *14*, 647.
[49] Neilson, J. C. (2009). *Vet. Med.*, *104*, 140-142, 144.
[50] Hussin, F., Aroua, M. K. & Daud, W. M. A. W. (2011). *Chem. Eng. J.*, *170*, 90.
[51] Darley, H. C. H. & Gray, G. R. (1988). Composition and properties of drilling and completion fluids, 5th ed. Butterworth-Heinemann, Woburn, Massachusetts.
[52] Rabah, A. A. & Abdelrahman, F. E. (2012). International Conference on chemical, civil and environment engineering (ICCEE 2012) 24-25 March, (Dubai).
[53] Srasra, E., Bergaya, F., Van Damme, H. & Riquib, N. K. A. (1989). *Appl. Clay Sci.*, *4*, 411.
[54] Siddiqui, M. K. H. (1968). Bleaching Earth, Pergamon Press, London.
[55] Clem, A. G. & Doehler, R. W. (1961). *Clays and Clay Minerals*, *10*, 272.
[56] Bulut, G., Chimeddorj, M., Esenli, F. & Çelik, M. S. (2009). *Appl. Clay Sci, 46*, 141.
[57] Langley, J. & Holroyd, D. (1990). US Patent No. 4, 913, 775.
[58] Takashima, M., Sano, & Ohara, S. (1993). *J. Imag. Sci. Tech.*, *37*, 163.
[59] Gates, W. P., Bouazza, A. & Churchman, G. J. (2009). *Elements*, *5*, 105.
[60] Milutinovic-Nikolic, A., Dostanic, J., Bankovic, P., Jovic-Jovicic, N., Lukic, S., Rosic, B. & Jovanovic, D. (2011). *J. Serb. Chem. Soc.*, *76*, 1411.

[61] Pinnavaia, T. J. (1983). *Science*, *220*, 365.
[62] Glu, T. A., Akkus, G., Onal, M. & Sarikaya, Y. (2003). *Turk. J. Chem.*, *27*, 675.
[63] Hamzah, N., Alias, A., Adnan, N. F., Nadzri, A. H., Nordin, N. M., Kassim, M. & Yarmo, M. A. (2011). *J. Sci. Tech.*, 59.
[64] Farshia, R., Belthura, C., Athreyasa, R.& Ramesh, G. J. (2013). *Int. J. Curr. Eng. Tech.*, 1, 172.
[65] Soetaredjo, F. E., Ayucitra, A., Ismadji, S. & Maukar, A. L. (2011). *Appl. Clay Sci.*, *53*, 341.
[66] Feng, J., Hu, X. & Yue, P. L. (2004). *Environ. Sci. Technol.*, *38*, 269.
[67] Kalbasi, R. J., Massah, A. R. & Daneshvarnejad, B. (2012). *Appl. Clay Sci.*, *55*, 1.
[68] Meleroa, J. A., Bautista, L. F., Iglesias, J., Morales, G. & Sanchez-Vazquez, R. (2014). *Appl Catal. B: Environmental*, *145*, 197.
[69] Aghabarari, B. & Dorostkar, N. (2014). *J. Taiwan Inst. Chem. Eng.*, *45*, 1468.
[70] Hefne, J. A., Mekhemer, W. K., Alandis, N. M., Aldayel, O. A. & Alajyan, T. (2010). *JKAU: Sci.*, *22*, 155.
[71] Al-Shahrani, S. S. (2012). *Int. J. Eng. Tech.*, *12*, 14.
[72] Hannachi, Y., Homri, T. & Boubaker, T. (2013). *The Hol. App. Env. 3*, 123.
[73] Weiss, K. & Bisson, L. (2002). *Am. J. Enol. Vitic.*, *53*, 28.
[74] Koyuncua, H., Kulb, A. R., Alimlic, A. C., Yildizc, N. & Ceylan, H. (2007). LWT, *40*, 489.
[75] Southern, L. L., Ward, T. L., Binder, T. D. & Hebert, L. G. (1994). *Poult. Sci.*, *73*, 848.
[76] Pasha, T. N., Mahmood, A., Khattak, F. M., Jabbar, M. A. & Khan, A. D. (2008). *Turk. J. Vet. Anim. Sci.*, *32*, 245.
[77] Liu, K., Wang, X. & Young, M. (2014). *J. Food Eng.*, *137*, 16.
[78] Komine, H. (2010). *Eng. Geo.*, *114*, 123.
[79] Wersin, P., Johnson, L. H. & McKinley, I. G. (2007). *Phy. Chem. Earth*, *32*, 780.
[80] Allen, C. C. & Wood, M. I. (1988). *Appl. Clay Sci.*, *3*, 11.
[81] Gomiz, E., Linares, J. & Delgado, R. (1992). *Appl. Clay Sci.*, *6*, 359.
[82] Bastedo, W. A. (1947). Pharmacology, therapeutics and prescription writing, 5th ed. Philadelphia: Saunders.
[83] Carraway, J. W., Kent, D., Young, K., Cole, A., Friedman, R. & Ward, K. R. (2008). *Resuscitation*, *78*, 230.
[84] Santos, M. F., Oliveira, C. M., Tachinski, C. T., Fernandes, M. P., Pich, C. T., Angioletto, E., Riella, H. G. & Fiori, M. A. (2011). *Int. J. Min. Proc.*, *100*, 51.
[85] Unsworth, E. F., Pearce, J., McMurray, C. H., Moss, B. W., Gordon, F. J. & Rice, D. (1989). *The Sci. Total Env.*, *85*, 339.
[86] Rebelo, M. & Viseras, C. (2011). A. Lopez-Galindo, F. Rocha and E. Ferreira da Silva, *App. Clay Sci.*, *51*, 258.
[87] Davies, T. C. (2013). *J. Afr. Earth Sci.*, *79*, 24.
[88] Onal, M. (2006). *Comm. Fac. Sci. Uni. Ank. Series B*, *52*, 7.
[89] Wanga, C., Jiang, X., Zhou, L., Xia, G., Chen, Z., Duan, M. & Jiang, X. (2013) *Chem. Eng. J.*, *219*, 469.
[90] Toor, M., Jin, B., S. Dai B. & Vimonses, V. (2014). *J. Ind. Eng. Chem.*, Preprint, at http://dx.doi.org/10.1016/j.jiec.2014.03.033.pdf.
[91] Williams, L. B., Haydel, S. E. & Ferrell, R. E. (2009). *Elements*, *5*, 99.

INDEX

A

access, 7, 96
accessibility, 25, 69
accounting, 13
acetic acid, 49, 50, 51, 143
acetone, 51
acid, 4, 5, 10, 11, 12, 13, 31, 34, 37, 39, 45, 49, 50, 60, 65, 71, 72, 76, 78, 79, 81, 82, 83, 84, 85, 88, 90, 91, 95, 102, 103, 104, 107, 110, 111, 112, 114, 119, 133, 143, 144, 145, 157, 163, 167, 168
acidic, 23, 37, 69, 70, 72, 105, 143, 157, 167
acidity, 34, 148, 157, 168
acne, 23
acrylate, 74
acrylic acid, 61, 74, 90
acrylonitrile, 10
activated carbon, 8, 27, 39, 89, 147
activation energy, 110
active site, 34, 69, 72
adaptability, 132
additives, 11, 94
adhesive properties, 21, 24, 168
adhesives, 97, 163, 168
adlayers, 76
adsorbent material, vii, 33, 117
adsorption isotherms, 7, 70, 82
advancement(s), viii, 134
adverse effects, 28, 37, 97, 139
aerosols, 97, 168
aflatoxin, 170
Africa, ix
aggregation, 52
agriculture, 39
aldehydes, 94
algae, 106
Algeria, 111
algorithm, 69
aliphatic hydrocarbon, vii
alkaline earth metals, 20
alkalinity, 37, 168
allergic reaction, 23
alternative energy, vii, 1
aluminium, 4, 5, 99, 100, 108, 131, 132, 133, 155
aluminum oxide, 142
amine, 26, 43, 49, 50, 51, 52, 60, 65, 72, 73, 83
amine group, 50, 51
amines, 13, 14, 26, 27, 30, 37, 87, 94, 163
amino, 37, 49, 60, 66, 72, 85, 115
amino groups, 37
ammonia, 5, 6, 104
ammonium, 26, 36, 43, 44, 46, 49, 51, 52, 53, 57, 60, 72, 74, 75, 76, 78, 82, 83, 90, 145, 157
ammonium salts, 43, 157
anemia, 97
aniline, 75, 82, 86
antibiotic, 171
antihistamines, 23
antioxidant, 23
anxiety, 23
aqueous solutions, 39, 44, 49, 72, 76, 84, 85, 86, 88, 89, 90, 91, 105, 106, 111, 112, 113, 114, 116, 117, 141, 142
aqueous suspension, 159
aromatic compounds, 37, 97
aromatic rings, 82
arsenic, 22, 97, 103, 108, 120
arsenic poisoning, 22
Asia, 149
assimilation, 19
atherosclerosis, 97
atmospheric pressure, 4
atoms, 21, 34, 35, 47, 72, 101, 115
autolysis, 38
automobile parts, 171
autopsy, 29
azo groups, 37

B

Bacillus subtilis, 106
bacteria, 19, 22, 23, 30, 95, 105, 106, 170
bactericides, 94
barriers, 4, 134, 167, 170
base, 4, 13, 90, 153, 170, 172
base catalysis, 90
baths, 24
batteries, vii, 1, 2, 8, 10, 11, 14, 96
behaviors, 90, 109, 110
Beijing, 137
bending, 44, 45, 46, 48, 51, 52
benefits, 4, 19, 108
benign, vii, 1, 14
benzene, 48, 49, 53, 60, 72, 94, 145
biodegradability, 142
biodiversity, 94
biomass, 105
biopolymer, 49, 117
bisphenol, 148
bleaching, 163
bleeding, 22
blends, 163
blood, 23, 29
blood stream, 29
bonding, 3, 23, 24, 34, 46, 76, 82, 155, 157, 163, 168
bonds, 39, 132, 160
borborygmi, 170
boreholes, 156
brass, 163
breakdown, 38, 97
building blocks, 20
burn, 95
by-products, 38, 39, 95, 143

C

Ca^{2+}, 3, 7, 13, 21, 36, 52, 99, 130, 133, 157, 160, 163, 166, 167
cables, 166
cadmium, 31, 96, 109, 113, 141
calcium, 3, 23, 28, 31, 51, 52, 100, 103, 129, 131, 132, 153, 154, 155, 157, 160, 165
calcium carbonate, 28, 157
cancer, 96, 97
candidates, vii, 1, 13
carbon, 4, 7, 10, 11, 12, 27, 39, 43, 57, 143
carbon atoms, 43
carbon dioxide (CO_2), 4, 12, 13, 14, 16
carbon materials, 10
carbon nanotubes, 12
carbonyl groups, 74
carboxyl, 37, 52, 73, 74, 83
carboxylic acid(s), 72, 143
carcinogenicity, 29
casting, 148, 163, 164, 165
catalyst, 5, 6, 12, 142, 144, 145, 146, 157, 160, 163, 168
catalytic activity, 169
catalytic effect, 5
catalytic system, 142, 143
cation, vii, 3, 13, 26, 27, 34, 35, 36, 40, 53, 70, 71, 74, 82, 83, 87, 102, 104, 105, 108, 111, 112, 113, 119, 129, 130, 132, 133, 134, 155, 160, 163, 169
cationic surfactants, 36, 44, 99, 101
C-C, 126
CEC, 34, 42, 44, 45, 46, 49, 59, 102, 113, 119
cellulose, 51, 91, 107, 167
central nervous system, 142
challenges, 1, 12, 39
chemical(s), vii, 2, 4, 5, 8, 12, 13, 19, 22, 24, 26, 31, 37, 38, 39, 41, 69, 70, 74, 84, 94, 95, 96, 102, 103, 105, 107, 108, 109, 113, 116, 132, 134, 140, 141, 155, 157, 158, 161, 163, 170, 172, 173
chemical characteristics, 161
chemical industry, 2
chemical interaction, 69, 170
chemical properties, vii, 19, 39, 155, 157, 163
chemical reactions, 8, 74
chemisorption, 54, 69, 110, 113, 114, 115
chicken, 31
children, 96
China, ix, 96, 121, 122, 153
chitin, 39, 117
chitosan, 39, 49, 50, 51, 60, 61, 67, 72, 73, 74, 86, 89, 90, 109, 115, 116, 117, 118, 142, 145, 146, 149
chlorination, 38
chloroform, 95
cholesterol, 23
chromatography, 90
chromium, 86, 96, 104, 106, 108, 114, 142
circulation, 25
classification, 2, 163
clay mineral, vii, 1, 2, 3, 4, 6, 7, 11, 19, 22, 26, 28, 33, 34, 35, 36, 39, 44, 85, 86, 87, 91, 100, 101, 113, 114, 129, 134, 139, 140, 141, 147, 153, 157, 166, 171
clay minerals, 1, 2, 3, 4, 6, 7, 19, 33, 34, 35, 36, 39, 85, 87, 91, 100, 101, 114, 129, 139, 140, 141, 171
clay swelling, 21
clean energy, vii, 1, 14
cleaning, 107, 157, 168
cleavage, 35, 101, 143

climate, 12
climate change, 12
clusters, 82
CMC, 50
C-N, 52
coal, 4, 7, 22, 38, 39, 96, 97
coatings, 135
cobalt, 6, 112
coffee, 106
coke, 140
color, 145, 166
combustion, 13, 38, 95, 96
commercial, 28, 103, 117, 118, 142, 145, 149, 155, 158, 169
common symptoms, 23
communication, 97
community(s), 30, 140
compatibility, 8, 99
competition, 111
competitiveness, 4
complexity, 164
composites, ix, 3, 11, 50, 51, 60, 65, 72, 73, 85, 91, 93, 102, 109, 117, 118, 140, 141, 142, 143, 149, 171
composition, 7, 13, 24, 36, 131, 132, 135, 155, 158, 160
compounds, 21, 39, 47, 76, 86, 94, 95, 111, 114, 129, 141, 143, 166, 168, 171
compressibility, 155
conductivity, vii, 8, 9, 10, 11, 12, 24, 129, 130, 132, 134, 167, 170
conductor, 8
conference, 87
configuration, 159
Congo, 57, 61, 73, 90, 146
conserving, 14
constituents, 6, 12
construction, vii, 19, 95
consumers, 139
consumption, 94, 140, 170
contact time, 86, 88, 111, 112, 114, 117
contaminant, 91, 96, 147
contaminated water, 39, 84, 109
contamination, 140, 147
COOH, 73, 74, 83
cooking, 169
cooling, 25
copper, 86, 97, 104, 108, 113, 115
cornea, 38
correlation, 71, 110, 111, 113, 115, 116
correlation coefficient, 110, 111, 113, 115, 116
corrosion, 130
cosmetic(s), 95, 98, 157, 163, 170, 171
cost, vii, 1, 4, 8, 12, 13, 14, 33, 39, 84, 87, 88, 93, 98, 103, 105, 107, 109, 118, 135, 139, 140, 141, 142, 145, 149, 157, 165, 171
cotton, 147
cough, 29, 96
covalent bond, 37, 115
covering, 39, 40
CPC, 44
critical analysis, 39
crops, 94, 147
crystal structure, 1, 153, 155
crystalline, 5, 29, 34, 100, 153
crystallinity, 98, 148
crystals, 33, 52
CSCE, 87
culture, 132
CV, 46, 146
cycles, 8, 11, 115, 147, 161
cycling, 5
Czech Republic, 28, 30

D

decomposition, 5, 142, 145, 153, 171
decontamination, 118, 149
Deer, 34, 86
defects, 97
deficiency(s), 28, 36
deformation, 46, 50
degradation, 37, 143, 144, 145, 146, 148, 167, 168
degradation rate, 145, 146
Department of Health and Human Services, 85
Department of the Interior, 30
deposition, 31, 100, 130
deposits, 19, 21, 154, 161
depression, 23
derivatives, 39, 88, 142, 144
dermatitis, 96, 97
desorption, 48, 89, 113, 114, 115, 116, 142, 146
destruction, 38
destructive process, 29
detergents, 94
detoxification, 28, 39, 83
developing nations, 118, 149
diarrhea, 28
dielectric constant, 9
diet, 28, 170
diffraction, 47, 50, 52
diffusion, 4, 8, 71, 73, 75, 83, 115, 117
digestibility, 169
digestion, 23
dimerization, 82
discharges, 95

diseases, 22, 23, 28, 31, 108, 140, 171
disinfection, 108
disorder, 47
dispersion, 5, 12, 27, 37, 45, 46, 48, 50, 51, 99, 133, 148, 157, 163
dissociation, 146
dissolved oxygen, 37
distilled water, 42, 45, 46, 47
distress, 97
distribution, 26, 38, 105, 154, 169
diversity, 37
DNA, 40, 97
DOI, 15, 16
doping, 145
dosage, 43, 70, 71, 117
drainage, 95, 102, 103, 162, 167
drawing, 24, 28
dressings, 170
drinking water, 93, 96, 97, 104, 107, 108, 118, 140, 149, 167
drying, 33, 44, 47, 136, 161, 163, 165
durability, 5, 12, 169
dyeing, 145
dyes, 33, 37, 38, 39, 40, 53, 54, 57, 58, 62, 70, 71, 72, 73, 74, 83, 85, 86, 87, 88, 89, 90, 94, 97, 133, 134, 140, 142, 145, 146, 147, 149, 167, 171

E

economic growth, 171
economic losses, 28
ecosystem, 140
eczema, 23
edema, 38
effluent(s), 38, 84, 86, 87, 94, 97, 108, 111, 135, 143, 144, 145, 149
Egypt, 149
elastomers, 98
electric field, 108
electrical conductivity, 11
electricity, 12, 108
electrodes, 8, 10, 11, 108
electrolyte, 8, 9, 10, 11, 12, 129, 159
electron(s), 34, 37, 70, 115
electroplating, 141
emission, 4
emulsion polymerization, 98, 146
emulsions, 26, 168
endocrine, 148
endothermic, 69, 71, 83, 111, 113, 116
endotoxemia, 30
endotoxins, 30
energy, vii, 1, 2, 3, 4, 6, 8, 12, 13, 14, 72, 98, 107, 108, 116, 157, 171
energy consumption, 108
energy density, 2, 4, 8
energy efficiency, 14
energy security, 4
energy supply, 6
engineering, viii, 97, 107, 134, 155, 163, 173
England, 86
entrapment, 26, 168
entropy, 114, 116
environment, vii, viii, 24, 29, 30, 34, 37, 38, 47, 94, 96, 97, 109, 120, 124, 139, 141, 142, 145, 170, 173
environmental issues, 29
Environmental Protection Agency (EPA), 96, 119, 120
environmental science, vii, viii
enzymes, 23
equilibrium, 40, 41, 69, 70, 71, 73, 75, 83, 85, 86, 90, 91, 112, 113, 116, 117, 141, 142
equipment, 164, 171
erosion, 161
ester, 169
ethanol, 48
ethyl alcohol, 48
ethylene, 8, 37, 102, 109
ethylene glycol, 37, 109
ethylene oxide, 8, 102
European Union (EU), 119
evidence, 22, 29
evolution, 130
experimental condition, 7, 40, 53, 74, 75, 104
exposure, 24, 29, 38, 96, 97, 142, 158
expulsion, 39
extraction, 155
extrusion, 163

F

families, 102
farmers, 28
farms, 30, 134
fat, 24
fatty acids, 94
fermentation, 169
ferric ion, 142
fertilizers, 94, 96, 135, 141, 163
fiber(s), 24, 37, 167
fibrosis, 97
films, 130, 146, 171
filtration, 107, 108, 166
financial support, 118

fire retardancy, 98
fish, 96
flocculation, 73, 146, 166
flooding, 168
flora and fauna, 94
fluid, 7, 24, 25, 157, 166, 167
food, 95, 153, 155, 169, 171, 172
food additive(s), 153, 172
food production, 155
force, 108
forest fire, 25, 170
formation, 43, 44, 45, 50, 71, 76, 83, 134, 135, 142, 143, 148, 167
formula, 36, 100, 129
fouling, 107
foundations, 167, 171
fractal dimension, 46
fractures, 24, 167
fragments, 28
France, 16, 100, 130
free energy, 112, 114, 116
FTIR, 44, 45, 52, 90, 117, 118
fuel cell, 2, 12, 14
functionalization, 109, 110
fungi, 22, 105, 106
fungus, 170
fusion, 3

G

gamma rays, 51
gangrene, 24
gel, 3, 8, 9, 22, 24, 25, 29, 39, 104, 153, 165, 167, 172
geology, 121, 173
Georgia, 85
Germany, 130, 153
glass transition temperature, 109
global warming, 12, 126
globalization, 30
glycerol, 168
glycol, 135
graphite, 10
grass, 147
gravimetric analysis, 118
gravity, 158
grazing, 169
Greece, 112, 153
green alga, 106
greenhouse gas(s), 13
grounding, 171
groundwater, vii, 19, 134, 147, 161, 167
Gulf of Mexico, 96

H

harmful effects, 24
hazardous substances, 38
healing, 22, 23, 24, 170, 171
health, 19, 22, 24, 28, 29, 97, 109, 120
health effects, 29, 120
health problems, 97
health status, 28
heart failure, 96
heavy metals, 24, 39, 40, 85, 89, 94, 103, 104, 105, 106, 107, 110, 111, 114, 123, 139, 140, 141, 149, 169
hemorrhage, 38
hemorrhoids, 22
herbicide, 147, 148
heterogeneity, 5
hexafluoropropylene, 8, 9
hexane, 49, 60, 66, 72
HFP, 8
highways, 166, 167
host, 44, 130, 131, 134, 161
human activity, 139
human body, 22, 23, 24
human health, 97, 171
humus, 104
Hungary, 96, 120
hybrid, 4, 8, 117, 127, 129, 146
hydrocarbons, 27, 95
hydrogels, 51
hydrogen, vii, 1, 4, 5, 12, 40, 76, 82, 108, 135, 142
hydrogen peroxide, 108, 142
hydrolysis, 115, 132
hydrophilic, vii, 25, 26, 27, 33, 36, 40, 43, 52, 70, 71, 99, 140, 163
hydrophilicity, 99, 117
hydrophobicity, 49
hydroquinone, 82, 91, 143
hydroxide, 102, 119, 132
hydroxyl, 34, 35, 51, 73, 79, 83, 108, 115, 133, 141, 142, 146
hyperfiltration, 134

I

immobilization, 49
immune system, 23
impact strength, 109
impregnation, 3
improvements, 14, 98
impurities, 24, 28, 29, 111, 139, 140, 157, 166
in vitro, 30, 171

in vivo, 171
India, ix, 33, 129, 139, 153, 155, 172
indium, 116
induction, 132
industrial chemicals, 94
industrial wastes, 141
industrialization, 140
industry(s), 21, 25, 26, 27, 28, 30, 31, 37, 39, 84, 89, 96, 97, 102, 132, 135, 140, 141, 142, 144, 145, 148, 163, 166, 172
infection, 22, 23
inflammation, 24
infrared spectroscopy, 118
ingestion, 29, 38, 97
ingredients, 94
inhibitor, 135
initial state, 131
injuries, 38
insecticide, 26, 169
insertion, 10, 11, 45, 99
insulation, 24
interaction process, 72
interface, 8, 26, 40, 48, 168
International Energy Agency (IEA), 16
interphase, 115
intoxication, 97
investment, 108
ion exchangers, 21, 104, 105, 123
ion-exchange, 13, 33, 89, 102, 104, 130, 134, 142
ionization, 39
ions, vii, 2, 8, 9, 10, 13, 21, 22, 26, 34, 35, 36, 42, 43, 44, 45, 47, 48, 53, 57, 71, 72, 75, 76, 82, 83, 86, 87, 90, 102, 104, 105, 106, 108, 109, 111, 112, 113, 114, 115, 116, 117, 140, 142, 146, 155, 156, 157, 159, 160, 165, 169
IR spectra, 48, 51, 52
IR spectroscopy, 112
iron, 20, 25, 98, 103, 111, 116, 130, 140, 143, 146, 154, 160, 163, 165
irradiation, 43, 113, 145
isolation, 30
isotherms, 40, 48, 76, 112, 117

J

Japan, 15, 130, 170
Jordan, 26, 30

K

K^+, 13, 36, 99, 160
kaolinite, vii, 2, 6, 7, 40, 84, 85, 89, 100, 110, 111, 141, 153
ketones, 94
kidney(s), 96, 97, 142
kill, 108
kinetic model, 69, 74, 113, 115, 118
kinetic studies, 112
kinetics, 41, 42, 53, 69, 70, 71, 72, 74, 75, 76, 83, 86, 87, 90, 104, 112, 114, 116, 117, 143, 148
KOH, 169
Korea, 139

L

laceration, 22
lakes, 100, 154, 161
land disposal, 141
landfills, 134, 167
lanthanum, 104
laws, 94
leaching, 163
lead, 87, 90, 96, 97, 99, 106, 109, 113, 114, 115, 134, 141, 142, 169
lesions, 30, 97
Lewis acids, 75
ligand, 40
light, 129, 130, 145, 154
lignin, 105
limestone, 102
liquid phase, 40
liquids, 26, 28, 30, 168
lithium, 8, 9, 10, 11
lithium insertion, 10, 11
liver, 23, 96
low temperatures, 158
lubricants, 21, 26, 95, 97, 171
lubricating oil, 142
lung cancer, 29
lung disease, 29
Luo, 86
lutetium, 104

M

macromolecules, 127
magnesium, 5, 102, 111, 117, 154, 155, 160
magnetic field, 25
mammals, 141
manganese, 141
manure, 134
mass, 5, 14, 21, 41, 44, 45, 52, 71, 73, 75, 111, 116, 131, 134

mass loss, 44, 52
materials, vii, viii, 1, 2, 3, 4, 5, 10, 11, 13, 14, 21, 22, 24, 26, 39, 52, 85, 99, 105, 108, 117, 118, 122, 127, 131, 132, 133, 140, 141, 149, 162, 165, 166, 168, 170, 171, 172, 173
materials science, viii
matrix, 9, 12, 74, 83, 98, 99, 109
matter, 6, 7, 19, 27, 94, 97, 108
maximum sorption, 114, 117
measurements, 6, 7, 43, 44, 117, 131
mechanical properties, 8, 38, 98, 109
media, 22, 26, 27, 89, 113, 134, 141, 171
medical, vii, 19, 30, 127, 153, 171, 172
medicine, 22, 31
Mediterranean, 140
membranes, 8, 9, 10, 12, 51, 73, 89, 102, 107, 108, 148
memory, 23
mercury, 96, 103, 109, 119
mesoporous materials, 3, 135
metabolism, 31
metal ion(s), 40, 89, 90, 93, 96, 98, 102, 104, 107, 108, 109, 111, 112, 113, 114, 115, 117, 140, 141
metal oxides, 5, 33, 141
metal recovery, 105
metals, 24, 37, 39, 94, 95, 96, 104, 107, 108, 109, 111, 112, 118, 133, 139, 140, 149, 160
meter, 132
methanol, 12
methyl group(s), 50, 53
methylene blue, 71, 87, 90, 91, 134
Mg^{2+}, 21, 35, 36, 133, 155, 157, 160, 163, 166, 167
mice, 30, 97
microcrystalline, 141
microorganisms, 98, 108
microscope, 34
microstructure(s), 10, 131, 132
microwave heating, 113
migration, 7, 12, 134
mineral resources, 171
mineralization, 145, 169
MIW, 102
mixing, 45, 48, 49, 51, 52, 141, 168
models, 40, 91, 113, 116, 118
modifications, 141, 157
modulus, 9, 109
moisture, 7, 165, 166, 167
moisture content, 7
molar ratios, 50, 73, 146
mold(s), 20, 163, 165
mole, 112, 143, 145
molecular oxygen, 142
molecular structure, 70, 115, 154
molecular weight, 27, 98
molecules, 4, 7, 21, 22, 28, 34, 35, 36, 37, 38, 39, 43, 45, 46, 47, 48, 49, 51, 52, 53, 57, 69, 70, 71, 72, 75, 76, 82, 101, 115, 134, 157, 159, 160
monolayer, 41, 44, 47, 48, 53, 76, 112, 117, 159
monomers, 10, 99
Montana, 19
montmorillinite, vii
Moon, 122
morphology, 8, 52, 98
mortality, 94
moulding, 163
mucosa, 28
multi-component systems, 84
multiwalled carbon nanotubes, 5
municipal solid waste, 38
mutations, 97

N

Na^{+}, 3, 13, 21, 26, 36, 45, 47, 48, 50, 52, 99, 115, 130, 133, 160, 163, 167
NaCl, 44, 47, 49, 52, 115
nanocomposites, ix, 8, 9, 10, 11, 12, 27, 49, 50, 51, 98, 99, 100, 102, 109, 115, 145, 146, 147, 148, 149, 169
nanocrystals, 141
nanofibers, 10, 99
nanometer scale, 98
nanoparticles, 4, 5, 98, 99, 140, 141, 142, 145, 147
nanostructures, 11
nanotechnology, 93, 99, 171
naphthalene, 75, 82, 97, 145
natural gas, 6
natural polymers, 109
nausea, 97
near infrared spectroscopy, 75
necrosis, 38
negative effects, 129
neodymium, 104
neural network, 69, 86
neurons, 69
neutral, 23, 88, 101, 102, 145, 154
NH_2, 50, 51, 71, 73, 83, 115
nickel, 5, 96, 97, 104, 105, 112, 115, 141, 154, 169
nitrates, 95
nitrobenzene, 75, 82
nitrogen, 47, 48, 49, 51, 72, 74, 145
nitrogen gas, 49
non-polar, vii, 33
North America, 30
NRF, 118
nucleation, 132

nuclei, 52
nucleic acid, 40, 91
nutraceutical, 167
nutrient(s), 19, 24, 28, 29, 31, 95, 105, 118, 148, 169
nutrition, 155

O

obstacles, 1
oceans, 161
octahedron-layer, vii
officials, 120
oil, 7, 26, 27, 120, 135, 139, 156, 157, 163, 166, 168, 169, 171
oil spill, 120
oleic acid, 169
oligomers, 99
olive oil, 144
one dimension, 98
operations, 27, 140
opportunities, 16
organ(s), 3, 9, 27, 31, 45, 86, 109, 115
organic chemicals, 26
organic compounds, vii, 3, 27, 37, 39, 76, 83, 91, 95, 101, 140, 166
organic matter, 6, 7, 19, 27, 34, 76, 108, 118, 129, 130, 147, 148, 157
organic pollutants, vii, 33, 34, 40, 77, 80, 139, 140, 171
organic polymers, 157
organic solvents, 21, 26, 94, 97, 115, 139, 140, 157
osmosis, 107, 134
osmotic pressure, 107
osteoporosis, 96
oxidation, 108, 143, 144, 146, 163, 166
oxidation products, 166
oxygen, 12, 21, 23, 28, 34, 35, 37, 74, 95, 130, 134
ozonation, 108
ozone, 108

P

PAA, 49, 107, 141
Pacific, 149
pain, 24, 96
paints, 26, 97, 157, 163, 171
Pakistan, 172
PAN, 9
parallel, 47, 52, 53
paralysis, 142
partition, 27, 75, 76, 82
passivation, 8
pathogens, 94, 95
pathways, 39, 105
peat, 28, 31, 39, 87
percolation, 28, 166
permeability, 12, 155, 156, 167, 171
permission, 101
peroxide, 108, 142, 144
personal hygiene, 95
pesticide, 26, 148
petroleum, 14, 27, 84, 95, 97, 121, 137, 141, 149, 160, 163
PGE, 8
pH, 23, 34, 39, 40, 45, 49, 53, 60, 61, 62, 63, 64, 65, 69, 70, 72, 74, 75, 77, 78, 79, 80, 81, 82, 83, 86, 111, 112, 114, 116, 117, 129, 130, 142, 143, 144, 145, 146, 147, 154, 160, 169
pharmaceutical(s), 31, 142, 155, 167
phase inversion, 148
phenol, 38, 39, 74, 75, 76, 77, 78, 79, 80, 82, 84, 85, 87, 88, 89, 91, 110, 142, 143, 144
phenol oxidation, 142, 144
phenolic compounds, 75, 76, 84, 89
phenolic resins, 38
Philadelphia, 174
phosphate(s), 94, 95, 96, 139, 141
phosphorus, 141
photocatalysts, 146
photodegradation, 145
photolysis, 146
photosynthesis, 37
physical interaction, 170
physical properties, vii, 1, 21, 99, 132, 135
physics, viii
pigs, 30
plants, 109, 139, 140
plasticity, 24, 33, 153, 155, 165, 168, 171
plastics, 169
platelets, 159, 160
platform, vii, 26
platinum, 88
plutonium, 104
pneumoconiosis, 29, 31
poison, 169
polar, 21, 34, 35, 36, 39, 73, 76, 82, 86, 157
polarity, 34, 115
pollutants, vii, 12, 33, 34, 37, 38, 39, 40, 77, 80, 93, 95, 102, 109, 133, 139, 140, 144, 149, 171
pollution, 93, 94, 95, 96
polyacrylamide, 60, 65, 72, 79, 81, 83, 85
polychlorinated biphenyls (PCBs), 95, 139, 140
polycyclic aromatic hydrocarbon, 139, 140

polymer, viii, 8, 9, 10, 12, 27, 48, 49, 74, 85, 86, 88, 93, 98, 99, 100, 102, 104, 109, 113, 127, 129, 130, 143, 148, 163
polymer chain(s), 74
polymer electrolytes, 9, 12
polymer matrix, 8, 99, 113
polymer nanocomposites, 99, 148, 163
polymer networks, 74
polymer synthesis, 143
polymeric materials, 3
polymeric matrices, 157
polymerization, 49, 51, 98
polymers, 10, 39, 99, 167
polypropylene, 109, 135
polysaccharide, 74, 83
polyvinyl acetate, 10
polyvinyl alcohol, 10
ponds, 134, 161, 167
population, 94
porosity, 4, 5, 7, 8, 10, 11, 113, 148, 157
porous materials, 3, 4
Portugal, 126
potassium, 49, 94, 103, 131, 154, 169
potato, 105
poultry, 31
power plants, 95
precipitation, 43, 102, 103, 104, 111, 116, 132, 140
preparation, 5, 10, 31, 33, 39, 40, 45, 46, 49, 50, 52, 87, 91, 99, 102, 113, 130, 145
preservation, 38
prevention, 22, 25, 28
principles, 173
producers, 155
prophylaxis, 31
propylene, 102
proteins, 40, 91, 160, 166
protons, 54, 157
Pseudomonas aeruginosa, 106
public health, 97
pulp, 38, 141, 144
purification, 22, 94, 153, 171
purity, 28, 163
PVP, 147
pyrite, 19, 103
pyrolysis, 10

Q

quality control, 171
quartz, 7, 19, 28, 33, 100, 110, 129, 153
quaternary ammonium, 36, 37, 43, 44, 45, 47, 70, 99
Queensland, 3, 173
quinones, 143

R

radiation, 22, 24, 84
radical mechanism, 143
radicals, 108, 142
radioactive isotopes, 24
radioactive waste, vii, 30, 94, 170, 172
radius, 13
raw materials, 25
reaction temperature, 43
reactions, 21, 26, 30, 33, 71, 87, 108, 141
reactivity, 21, 98, 132
reading, 133
reagents, 102, 103, 108, 112
recovery, 23, 84, 107, 108, 116
recycling, 94
red mud, 39
regeneration, 13, 84, 115
Registry, 85
regression, 117
rejection, 108
reliability, 108
remediation, vii, 88
renewable energy, 2, 14
renewable energy technologies, 2
replication, 97
repulsion, 74, 83, 148
requirements, 25, 107
reserves, 1
resins, 39, 88, 104, 122, 141, 142
resistance, 8, 31, 38, 98, 132
resolution, 34
resources, 6, 30, 121, 161
reusability, 135
reverse osmosis, 140
rings, 37, 82, 145
risk(s), 6, 145
RNA, 40
Romania, 113
room temperature, 5, 8, 9, 42, 43, 45, 47, 49, 50, 51
runoff, 95
Russia, 153

S

salinity, 131
salmonella, 95
salt formation, 131
salt production, 108
salts, 37, 72, 87, 111, 116, 131, 154
samarium, 104
saturation, 25, 157

sawdust, 105
SAXS, 85
science, vii, viii, 2, 87
seafood, 97
sediment, 28, 95
sedimentation, 29, 116
seed, 105, 135
selectivity, vii, 33, 39, 98, 105, 111, 114, 171
selenium, 103, 117, 142
semi-permeable membrane, 107
sensors, 131
sewage, 94, 95, 96, 140, 141, 148, 167, 168
shape, 5, 82
shear, 133, 155, 163
shelf life, 166, 167
showing, 20, 44, 47, 115, 132, 144
shrimp, 105
silane, 109
silanol groups, 101
silica, 2, 3, 4, 20, 29, 35, 39, 45, 46, 48, 90, 101, 117, 153, 155, 163, 167
silicon, 160
silicosis, 29
silver, 108, 169
Singapore, 123, 126
SiO_2, 36, 135, 158
SiO_2 surface, 135
skin, 23, 24, 37, 38, 96, 97, 148, 170
sludge, 29, 39, 96, 105, 120, 141
smooth muscle, 31
sodium, 3, 6, 19, 20, 26, 28, 31, 44, 51, 94, 100, 102, 103, 105, 108, 114, 129, 131, 132, 134, 148, 153, 154, 155, 156, 157, 159, 160, 168, 169
sodium hydroxide, 102, 114
soil particles, 168
solid phase, 40
solubility, 27, 39, 76, 83, 166
solvation, 157
solvent molecules, 38
solvents, 95, 115, 142
sorption, 3, 7, 84, 87, 89, 90, 91, 111, 112, 114, 116, 118, 140, 142, 146, 160, 169, 170
sorption materials, 169
sorption process, 111, 112, 114, 160
South Africa, ix, 1, 93, 118
Spain, 130
species, 4, 29, 37, 43, 69, 70, 72, 75, 76, 83, 101, 104, 106, 107, 110, 111, 134, 141, 163, 170
specific gravity, 131
specific surface, vii, 3, 4, 5, 10, 11, 13, 44, 47, 49, 111, 129, 130, 148, 160
sponge, 22
sputum, 29
stability, 3, 5, 10, 12, 26, 129, 135, 142, 163, 168
stabilization, 8, 168
stabilizers, 26, 94
starch, 109
state(s), 19, 21, 24, 96, 108, 130, 132, 155, 167, 170
steel, 130, 145, 165
stimulation, 132
stoichiometry, 13
stomach, 28, 31, 97
storage, vii, 1, 2, 3, 4, 5, 6, 7, 8, 14, 97, 134, 141
storage media, 2
stress, 133
stretching, 42, 44, 45, 46, 48, 49, 50, 51, 52, 53, 75
structure, vii, 7, 10, 13, 22, 33, 34, 35, 36, 45, 46, 50, 51, 53, 69, 70, 74, 76, 87, 91, 99, 101, 115, 129, 130, 133, 134, 141, 155, 157, 158, 159, 170
styrene, 144, 147
substitution(s), 26, 35, 36, 101, 114, 134, 155, 160
substrate(s), 10
sucrose, 10, 11
sulfate, 103, 139
sulfur, 95, 142, 145
sulfur dioxide, 95
sulfuric acid, 169
Sun, 37, 88, 90, 152
supplementation, 28
surface area, 3, 4, 5, 7, 8, 10, 11, 13, 21, 24, 25, 26, 34, 38, 43, 44, 45, 46, 47, 48, 49, 50, 53, 72, 75, 93, 102, 110, 113, 115, 132, 140, 141, 145, 157, 160, 163, 166, 168, 169, 170
surface energy, 3, 44
surface mining, 161
surface modification, 26, 27, 99, 156
surface properties, 101, 171
surface tension, 39
surface treatment, 98
surfactant(s), 3, 39, 42, 43, 44, 45, 46, 47, 48, 51, 52, 57, 61, 68, 69, 71, 74, 75, 76, 79, 81, 82, 83, 84, 87, 88, 89, 90, 91, 94, 101, 105, 107, 113, 135, 147
suspensions, 21, 25, 44, 88, 157, 166, 170
sustainability, 4
Sweden, 149
swelling, 2, 3, 20, 21, 24, 100, 102, 130, 131, 132, 134, 135, 153, 155, 156, 157, 159, 160, 161, 167, 170, 171
swelling process, 134
symmetry, 35, 169
symptoms, 23
synergistic effect, 73
synthesis, 10, 27, 43, 46, 51, 84, 96, 100, 109, 116, 135, 142, 143

T

Taiwan, 120, 174
talc, 2
tanks, 4, 97
target, 26
technological advancement, 94
technological developments, 2
technology(s), vii, 2, 4, 13, 14, 16, 22, 38, 98, 102, 107, 108, 140, 145, 147, 153, 171
TEM, 44, 147
temperature, 4, 5, 8, 10, 12, 34, 39, 40, 52, 69, 70, 71, 74, 75, 76, 82, 83, 109, 110, 112, 113, 116, 141, 142, 146, 147, 154, 169, 171
tensile strength, 109
testing, 131, 161
tetrahedron-layer, vii
textiles, 96, 97
texture, 11, 13, 161
therapeutics, 174
thermal analysis, 118
thermal conductivity, vii
thermal decomposition, 5, 109
thermal degradation, 52, 163
thermal energy, 2, 3
thermal stability, 8, 12, 98, 109, 147
thermal treatment, 13, 168
thermodynamic parameters, 86
thermodynamics, 88
thermoplastics, 98
thermosets, 98
thorium, 104
tibia, 31
titanium, 98
toluene, 49, 115
total energy, 70
total product, 142
toxic metals, 94, 112, 140
toxic substances, 38, 140
toxicity, 29, 37, 120, 142
toxin, 22
transformation, 100, 113
transition metal, 8
transmission, 163
transport, 91, 134, 167
transportation, 12
treatment, vii, 2, 10, 13, 19, 22, 24, 37, 38, 39, 45, 72, 85, 87, 90, 93, 94, 98, 102, 104, 105, 107, 108, 109, 110, 113, 130, 135, 139, 140, 145, 160, 169, 170
treatment methods, 105
triggers, 130
Turkey, 141, 153

U

U.S. Geological Survey, 30, 121
UK, 101
Ukraine, 153
ultrasound, 146
UN, 119
United States (USA), 30, 96, 100, 121, 153
uranium, 104
urea, 147
UV irradiation, 108, 143, 147
UV light, 145
UV-irradiation, 108

V

valence, 70
variables, 93
variations, 7, 27, 35, 53, 148, 158
vegetable oil, 163
vegetation, 25, 26, 161, 169, 170
vibration, 42, 45, 50, 51, 52
victims, 170
vinylidene fluoride, 9
viruses, 22, 95
viscosity, 21, 25, 157, 159, 168, 170
vitamins, 29
volcanic ash, vii, 1, 28, 100, 132, 135, 153, 161, 171
vomiting, 97

W

Washington, 121
waste, vii, 3, 27, 39, 72, 85, 89, 95, 96, 103, 106, 108, 122, 130, 134, 140, 168, 169, 170
waste heat, 3
waste management, vii
wastewater, 22, 27, 37, 39, 72, 83, 84, 86, 87, 89, 90, 91, 93, 94, 95, 96, 98, 102, 103, 104, 105, 106, 107, 108, 109, 112, 117, 122, 134, 139, 140, 145, 148, 153, 163, 169, 171
water absorption, 129
water permeability, 167
water purification, 33, 36, 98, 139
water quality, 91, 119, 145
water sorption, 90
water supplies, 94
water vapor, 3
waterways, 96
wave number, 42
WAXS, 85
weight gain, 28

weight loss, 24
weight ratio, 4
wells, 134, 156
wetting, 136, 161
wildlife, 161
wood, 38, 39workers, 3, 29, 50, 74, 109
World Health Organization (WHO), 38, 91, 96, 119, 142

X

X-ray diffraction (XRD), 43, 44, 46, 47, 48, 50, 51, 52, 75, 110, 117, 119, 132

Y

yeast, 105
yield, 133
ytterbium, 104

Z

zeolites, 3, 39, 89, 90, 104, 105, 129, 141, 169
zinc, 97, 130, 154
zirconium, 114
ZnO, 130, 143